THE SCIENCE OF
DISCWORLD

HBP

THE SCIENCE OF
DISCWORLD

TERRY PRATCHETT
IAN STEWART & JACK COHEN

EBURY
PRESS

First published in Great Britain in 1999

This revised edition published in 2002 by Ebury Press

9 10 8

EBURY PRESS
Random House, 20 Vauxhall Bridge Road, London SW1V 2SA

RANDOM HOUSE AUSTRALIA PTY LIMITED
20 Alfred Street, Milsons Point, Sydney, New South Wales 2061, Australia

RANDOM HOUSE NEW ZEALAND LIMITED
18 Poland Road, Glenfield, Auckland 10, New Zealand

RANDOM HOUSE (PTY) LIMITED
Isle of Houghton, Corner of Boundary Road & Carse O'Gowrie,
Houghton 2198, South Africa

The Random House Group Limited Reg. No. 954009

www.randomhouse.co.uk

Papers used by Ebury Press are natural, recyclable products made from
wood grown in sustainable forests.

A CIP catalogue record for this book is available from the
British Library.

ISBN 9780091886578 (from Jan 2007)
ISBN 0 09 188657 0

Printed and bound in Great Britain by Bookmarque Ltd, Croydon, Surrey

*'Any sufficiently advanced technology
is indistinguishable from magic'*
ARTHUR C. CLARKE

---

*'Any technology distinguishable from magic
is insufficiently advanced'*
GREGORY BENFORD

---

*'The reason why truth is so much stranger
than fiction is that there is no requirement
for it to be consistent.'*
MARK TWAIN

'There are no turtles anywhere'
PONDER STIBBONS

CONTENTS

THE STORY
STARTS HERE ...

ONCE UPON A TIME, there was Discworld. There still is an adequate supply.

Discworld is the flat world, carried through space on the back of a giant turtle, which has been the source of – so far – twenty-seven novels, four maps, an encylopaedia, two animated series, t-shirts, scarves, models, badges, beer, embroidery, pens, posters, and probably, by the time this is published, talcum power and body splash (if not, it can only be a matter of time).

It has, in short, become immensely popular.

And Discworld runs on magic.

Roundworld – our home planet, and by extension the universe in which it sits – runs on rules. In fact, it simply *runs*. But we have watched the running, and those observations and the ensuing deductions are the very basis of science.

Magicians and scientists are, on the face of it, poles apart. Certainly, a group of people who often dress strangely, live in a world of their own, speak a specialized language and frequently make statements that appear to be in flagrant breach of common sense have *nothing* in common with a group of people who often dress strangely, speak a specialized language, live in ... er ...

Perhaps we should try this another way. Is there a connection between magic and science? Can the magic of Discworld, with its eccentric wizards, down-to-Earth witches, obstinate trolls, fire-breathing dragons, talking dogs, and personified DEATH, shed any useful light on hard, rational, solid, Earthly science?

We think so.

We'll explain why in a moment, but first, let's make it clear what *The Science of Discworld* is *not*. There have been several media tie-in *The Science of* ... books, such as *The Science of the X-Files* and *The Physics of Star Trek*. They tell you about areas of today's science that may one day lead to the events or devices that the fiction depicts. Did aliens crash-land at Roswell? Could an anti-matter warp drive ever be invented? Could we ever have the ultra long-life batteries that Scully and Mulder must be using in those torches of theirs?

We could have taken that approach. We could, for example, have pointed out that Darwin's theory of evolution explains how lower lifeforms can evolve into higher ones, which in turn makes it entirely reasonable that a human should evolve into an orangutan (while remaining a librarian, since there is no higher life form than a librarian). We could have speculated on which DNA sequence might reliably incorporate asbestos linings into the insides of dragons. We might even have attempted to explain how you could get a turtle ten thousand miles long.

We decided not to do these things, for a good reason ... um, two reasons.

The first is that it would be ... er ... dumb.

And this because of the second reason. Discworld does not run on scientific lines. Why pretend that it might? Dragons don't breathe fire because they've got asbestos lungs – they breathe fire because everyone knows that's what dragons *do*.

What runs Discworld is deeper than mere magic and more powerful than pallid science. It is *narrative imperative*, the power of story. It plays a role similar to that substance known as phlogiston, once believed to be that principle or substance within inflammable things that enabled them to burn. In the Discworld universe, then, there is narrativium. It is part of the spin of every atom, the drift of every cloud. It is what causes them to be what they are and continue to exist and take part in the ongoing story of the world.

On Roundworld, things happen because the *things* want to happen.* What people want does not greatly figure in the scheme of things, and the universe isn't there to tell a story.

With magic, you can turn a frog into a prince. With science, you

---

* In a manner of speaking. They happen because things obey the rules of the universe. A rock has no detectable opinion about gravity.

can turn a frog into a Ph.D and you still have the frog you started with.

That's the conventional view of Roundworld science. It misses a lot of what actually makes science tick. Science doesn't just exist in the abstract. You could grind the universe into its component particles without finding a single trace of Science. Science is a structure created and maintained by people. And people choose what interests them, and what they consider to be significant and, quite often, they have thought narratively.

Narrativium is powerful stuff. We have always had a drive to paint stories on to the Universe. When humans first looked at the stars, which are great flaming suns an unimaginable distance away, they saw in amongst them giant bulls, dragons, and local heroes.

This human trait doesn't affect what the rules *say* – not much, anyway – but it does determine which rules we are willing to contemplate in the first place. Moreover, the rules of the universe have to be able to produce everything that we humans observe, which introduces a kind of narrative imperative into science too. Humans think in stories*. Classically, at least, science itself has been the discovery of 'stories' – think of all those books that had titles like *The Story of Mankind*, *The Descent of Man*, and, if it comes to that, *A Brief History of Time*.

Over and above the stories of science, though, Discworld can play an even more important role: What if? We can use Discworld for thought experiments about what science might have looked like if the universe had been different, or if the history of science had followed a different route. We can look at science from the outside.

To a scientist, a thought experiment is an argument that you can run through in your head, after which you understand what's going on so well that there's no need to do a real experiment, which is of course a great saving in time and money and prevents you from getting embarrassingly inconvenient results. Discworld takes a more practical view – there, a thought experiment is one that you can't do and which wouldn't work if you could. But the kind of thought

---

* It took three years for this sentence to sink in. When it did, we wrote *The Science of Discworld II: The Globe*.

experiment we have in mind is one that scientists carry out all the time, usually without realizing it; and you don't need to do it, because the whole *point* is that it wouldn't work. Many of the most important questions in science, and about our understanding of it, are *not* about how the universe actually is. They are about what would happen if the universe were different.

Someone asks 'why do zebras form herds?' You could answer this by an analysis of zebra sociology, psychology, and so on … or you could ask a question of a very different kind: 'What would happen if they didn't?' One fairly obvious answer to that is 'They'd be much more likely to get eaten by lions.' This immediately suggests that zebras form herds for self-protection – and now we've got some insight into what zebras actually do by contemplating, for a moment, the possibility that they might have done something else.

Another, more serious example is the question 'Is the solar system stable?', which means 'Could it change dramatically as a result of some tiny disturbance?' In 1887 King Oscar II of Sweden offered a prize of 2,500 crowns for the answer. It took about a century for the world's mathematicians to come up with a definite answer: 'Maybe'. (It was a good answer, but they didn't get paid. The prize had already been awarded to someone who didn't get the answer and whose prizewinning article had a big mistake right at the most interesting part. But when he put it right, at his own expense, he invented Chaos Theory and paved the way for the 'maybe'. Sometimes, the best answer is a more interesting question.) The point here is that stability is not about what a system is actually doing: it is about how the system would change if you disturbed it. Stability, by definition, deals with 'what if?'.

Because a lot of science is really about this non-existent world of thought experiments, our understanding of science *must* concern itself with worlds of the imagination as well as with worlds of reality. Imagination, rather than mere intelligence, is the truly human quality. And what better world of the imagination to start from than Discworld? Discworld is a consistent, well-developed universe with its own kinds of rules, and convincingly real people live on it despite the substantial differences between their universe's rules and ours. Many of them also have a thoroughgoing grounding in 'common sense', one of science's natural enemies.

Appearing regularly within the Discworld canon are the buildings and faculty of Unseen University, the Discworld's premier college of magic. The wizards* are a lively bunch, always ready to open any door that has 'This door to be kept shut' written on it or pick up anything that has just started to fizz. It seemed to us that they could be useful ...

Clearly, as the wizards of Unseen University believe, this world is a parody of the Discworld one. If we, or they, compare Discworld's magic to Roundworld science, the more similarities and parallels we find. And when we didn't discover parallels, we found that the *differences* were very revealing. Science takes on a new character when you stop asking questions like 'What does newt DNA look like?' and instead ask 'I wonder how the wizards would react to this way of thinking about newts?'

There is no science *as such* on Discworld. So we have put some there. By magical means, the wizards on Discworld are led to create their own brand of science – some kind of 'pocket universe' in which magic no longer works, but rules do. Then, as the wizards learn to understand how the rules make interesting things happen – rocks, bacteria, civilizations – we watch them watching ... well, us. It's a sort of recursive thought experiment, or a Russian doll wherein the smaller dolls are opened up to find the largest doll inside.

And then we found that ... ah, but that is *another* story.

TP, IS, & JC, DECEMBER 1998

PS We have, we are afraid, mentioned in the ensuing pages Schrödinger's Cat, the Twins Paradox, and that bit about shining a torch ahead of a spaceship travelling at the speed of light. This is because, under the rules of the Guild of Science Writers, they *have*

---

* Like the denizens of any Roundworld university, they have unlimited time for research, unlimited funds and no worries about tenure. They are also by turns erratic, inventively malicious, resistant to new ideas until they've become old ideas, highly creative at odd moments and perpetually argumentative – in this respect they bear no relation to their Roundworld counterparts at *all*.

to be included. We have, however, tried to keep them short.

We've managed to be very, very brief about the Trousers of Time, as well.

PPS Sometimes scientists change their minds. New developments cause a rethink. If this bothers you, consider how much damage is being done to the world by people for whom new developments do *not* cause a rethink.

This second edition has been changed to reflect three years of scientific progress... forwards or backwards. (You will find both.) And we've added two completely new chapters: one on the life of dinosaurs, because the existing chapter on the death of dinosaurs seemed a bit depressing; and one on cosmic disasters, because in many ways the universe *is* depressing.

The Discworld story has proved more robust than the science. As should be expected. Discworld makes so much more sense than Roundworld does.

TP, IS, & JC, JANUARY 2002

ONE

SPLITTING THE THAUM

 SOME QUESTIONS SHOULD NOT BE ASKED. However, someone always does.

'How does it work?' said Archchancellor Mustrum Ridcully, the Master of Unseen University.

This was the kind of question that Ponder Stibbons hated almost as much as 'How much will it cost?' They were two of the hardest questions a researcher ever had to face. As the university's *de facto* head of magical development, he especially tried to avoid questions of finance at all costs.

'In quite a complex way,' he ventured at last.

'Ah.'

'What *I'd* like to know,' said the Senior Wrangler, 'is when we're going to get the squash court back.'

'You never play, Senior Wrangler,' said Ridcully, looking up at the towering black construction that now occupied the centre of the old university court.*

'I might want to one day. It'll be damn hard with that thing in the way, that's my point. We'll have to completely rewrite the rules.'

Outside, snow piled up against the high windows. This was turning out to be the longest winter in living memory – so long, in fact, that living memory itself was being shortened as some of the older citizens succumbed. The cold had penetrated even the thick and

---

* Wizard or 'Real' Squash bears very little relationship to the high speed sweaty hath playcd elsewhere. Wizards see no point in moving fast. The ball is lobbed lazily. Certain magical inconsistencies are built into the floor and walls, however, so that the wall a ball hits is not necessarily the wall it rebounds from. This was one of the factors which, Ponder Stibbons realized some time afterwards, he really ought to have taken into consideration. Nothing excites a magical particle like meeting itself coming the other way.

15

ancient walls of Unseen University itself, to the general concern and annoyance of the faculty. Wizards can put up with any amount of deprivation and discomfort, provided it is not happening to them.

And so, at long last, Ponder Stibbons's project had been authorized. He'd been waiting three years for it. His plea that splitting the thaum would push back the boundaries of human knowledge had fallen on deaf ears; the wizards considered that pushing back the boundaries of *anything* was akin to lifting up a very large, damp stone. His assertion that splitting the thaum might significantly increase the sum total of human happiness met with the rejoinder that everyone seemed pretty happy enough already.

Finally he'd ventured that splitting the thaum would produce vast amounts of raw magic that could very easily be converted into cheap heat. That worked. The Faculty were lukewarm on the subject of knowledge for knowledge's sake, but they were boiling hot on the subject of warm bedrooms.

Now the other senior wizards wandered around the suddenly-cramped court, prodding the new thing. Their Archchancellor took out his pipe and absent-mindedly knocked out the ashes on its matt black side.

'Um ... please don't do that, sir,' said Ponder.

'Why not?'

'There might be ... it might ... there's a chance that ...' Ponder stopped. 'It will make the place untidy, sir,' he said.

'Ah. Good point. So it's not that the whole thing might explode, then?'

'Er ... no, sir. Haha,' said Ponder miserably. 'It'd take a lot more than that, sir –'

There was a *whack* as a squash ball ricocheted off the wall, rebounded off the casing, and knocked the Archchancellor's pipe out of his mouth.

'That was *you*, Dean,' said Ridcully accusingly. 'Honestly, you fellows haven't taken any notice of this place in years and suddenly you all want to – Mr Stibbons? Mr Stibbons?'

He nudged the small mound that was the hunched figure of the

University's chief research wizard. Ponder Stibbons uncurled slightly and peered between his fingers.

'I really think it might be a *good* idea if they stopped playing squash, sir,' he whispered.

'Me too. There's nothing worse than a sweaty wizard. Stop it, you fellows. And gather round. Mr Stibbons is going to do his presentation.' The Archchancellor gave Ponder Stibbons a rather sharp look. 'It is going to be very informative and interesting, isn't it, Mister Stibbons. He's going to tell us what he spent AM$55,879.45p on.'

'And why he's ruined a perfectly good squash court,' said the Senior Wrangler, tapping the side of the thing with his squash racket.

'And if this is *safe*,' said the Dean. 'I'm against dabbling in physics.'

Ponder Stibbons winced.

'I assure you, Dean, that the chances of anyone being killed by the, er, reacting engine are even greater than the chance of being knocked down while crossing the street,' he said.

'Really? Oh, well ... all right then.'

Ponder reconsidered the impromptu sentence he'd just uttered and decided, in the circumstances, not to correct it. Talking to the senior wizards was like building a house of cards; if you got *anything* to stay upright, you just breathed out gently and moved on.

Ponder had invented a little system he'd called, in the privacy of his head, Lies-to-Wizards. It was for their own good, he told himself. There was no *point* in telling your bosses *everything*; they were busy men, they didn't want *explanations*. There was no *point* in burdening them. What they wanted was little stories that they felt they could understand, and then they'd go away and stop worrying.

He'd got his students to set up a small display at the far end of the squash court. Beside it, with pipes looping away through the wall into the High Energy Magic building next door, was a terminal to Hex, the University's thinking engine. And beside that was a plinth on which was a very large red lever, around which someone had tied a pink ribbon.

Ponder looked at his notes, and then surveyed the faculty.

'Ahem …' he began.

'I've got a throat sweet somewhere,' said the Senior Wrangler, patting his pockets.

Ponder looked at his notes again, and a horrible sense of hopelessness overcame him. He realized that he could explain thaumic fission very well, provided that the person listening already knew all about it. With the senior wizards, though, he'd need to explain the meaning of every word. In some cases this would mean words like 'the' and 'and'.

He glanced down at the water jug on his lectern, and decided to extemporize.

Ponder held up a glass of water.

'Do you realize, gentlemen,' he said, 'that the thaumic potential in this water … that is, I mean to say, the magical field generated by its narrativium content which tells it that it *is* water and lets it keep on being water instead of, haha, a pigeon or a frog … would, if we could release it, be enough to move this whole university all the way to the moon?'

He beamed at them.

'Better leave it in there, then,' said the Chair of Indefinite Studies.

Ponder's smile froze.

'Obviously we cannot extract *all* of it,' he said, 'But we –'

'Enough to get a small part of the university to the moon?' said the Lecturer in Recent Runes.

'The Dean could do with a holiday,' said the Archchancellor.

'I resent that remark, Archchancellor.'

'Just trying to lighten the mood, Dean.'

'*But we can* release just enough for all kinds of useful work,' said Ponder, already struggling.

'Like heating my study,' said the Lecturer in Recent Runes. 'My water jug was iced up *again* this morning.'

'Exactly!' said Ponder, striking out madly for a useful Lie-to-Wizards. 'We can use it to boil a great big kettle! That's all it is! It's perfectly harmless! Not dangerous in any way! That's why the

University Council let me build it! You wouldn't have let me build it if it was dangerous, would you?'

He gulped down the water.

As one man, the assembled wizards took several steps backwards.

'Let us know what it's like up there,' said the Dean.

'Bring us back some rocks. Or something,' said the Lecturer in Recent Runes.

'Wave to us', said the Senior Wrangler. 'We've got quite a good telescope.'

Ponder stared at the empty glass, and readjusted his mental sights once more.

'Er, no,' he said. 'The fuel has to go inside the reacting engine, you see. And then ... and then ...'

He gave up.

'The magic goes round and round and it comes up under the boiler that we have plumbed in and the university will then be lovely and warm,' he said. 'Any questions?'

'Where does the coal go?' said the Dean. 'It's wicked what the dwarfs are charging these days.'

'No, sir. No coal. The heat is ... free,' said Ponder. A little bead of sweat ran down his face.

'Really?' said the Dean. 'That'll be a saving, then, eh, Bursar? Eh? Where's the Bursar?'

'Ah ... er ... the Bursar is assisting me today, sir,' said Ponder. He pointed to the high gallery over the court. The Bursar was standing there, smiling his distant smile, and holding an axe. A rope was tied around the handrail, looped over a beam, and held a long heavy rod suspended over the centre of the reaction engine.

'It is ... er ... *just* possible that the engine may produce too much magic,' said Ponder. 'The rod is lead, laminated with rowan wood. Together they naturally damp down any magical reaction, you see. So if things get too ..., if we want to settle things down, you see, he just chops through the rope and it drops into the very centre of the reacting engine, you see.'

'What's that man standing next to him for?'

'That's Mr Turnipseed, my assistant. He's the backup fail-safe device.'

'What does he do, then?'

'His job is to shout "For gods' sakes cut the rope now!" sir.'

The wizard nodded at one another. By the standards of Ankh-Morpork, where the common thumb was used as a temperature measuring device, this was health and safety at work taken to extremes.

'Well, that all seems safe enough to me,' said the Senior Wrangler.

'Where did you get the idea for this, Mister Stibbons?' said Ridcully.

'Well, er, a lot of it is from my own research, but I got quite a few leads from a careful reading of the Scrolls of Loko in the Library, sir.' Ponder reckoned he was safe enough there. The wizards liked ancient wisdom, provided it was ancient *enough*. They felt wisdom was like wine, and got better the longer it was left alone. Something that hadn't been known for a few hundred years probably wasn't worth knowing.

'Loko ... Loko ... Loko,' mused Ridcully. 'That's up on Uberwald, isn't it?'

'That's right, sir.'

'Tryin' to bring it to mind,' Ridcully went on, rubbing his beard. 'Isn't that where there's that big deep valley with the ring of mountains round it? Very deep valley indeed, as I recall.'

'That's right, sir. According to the library catalogue the scrolls were found in a cave by the Crustley Expedition –'

'Lots of centaurs and fauns and other curiously shaped magical whatnots are there, I remember reading.'

'Is there, sir?'

'Wasn't Stanmer Crustley the one who died of planets?'

'I'm not familiar with –'

'Extremely rare magical disease, I believe.'

'Indeed, sir, but –'

'Now I come to think about it, everyone on that expedition contracted something seriously magical within a few months of getting back,' Ridcully went on.

'Er, yes, sir. The suggestion was that there was some kind of curse on the place. Ridiculous notion, of course.'

'I somehow feel I need to ask, Mister Stibbons ... what chance is there of this just blowin' up and destroyin' the entire university?'

Ponder's heart sank. He mentally scanned the sentence, and took refuge in truth. 'None, sir.'

'Now try honesty, Mister Stibbons.' And that was the problem with the Archchancellor. He mostly strode around the place shouting at people, but when he did bother to get all his brain cells lined up he could point them straight at the nearest weak spot.

'Well ... in the unlikely event of it going seriously wrong, it ... wouldn't *just* blow up the university, sir.'

'What would it blow up, pray?'

'Er ... everything, sir.'

'Everything there is, you mean?'

'Within a radius of about fifty thousand miles out into space, sir, yes. According to HEX it'd happen instantaneously. We wouldn't even know about it.'

'And the odds of this are ... ?'

'About fifty to one, sir.'

The wizards relaxed.

'That's pretty safe. I wouldn't bet on a horse at those odds,' said the Senior Wrangler. There was half an inch of ice on the *inside* of his bedroom windows. Things like this give you a very personal view of risk.

TWO
SQUASH COURT SCIENCE

 A SQUASH COURT CAN BE USED to make things go *much* faster than a small rubber ball ...

On 2 December 1942, in a squash court in the basement of Stagg Field at the University of Chicago, a new technological era came into being. It was a technology born of war, yet one of its consequences was to make war so terrible in prospect that, slowly and hesitantly, war on a global scale became less and less likely.* At Stagg Field, the Roman-born physicist Enrico Fermi and his team of scientists achieved the world's first self-sustaining nuclear chain reaction. From it came the atomic bomb, and later, civilian nuclear power. But there was a far more significant consequence: the dawn of Big Science and a new style of technological change.

Nobody played squash in the basement of Stagg Field, not while the reactor was in place – but a lot of the people working in the squash court had the same attitudes as Ponder Stibbons ... mostly insatiable curiosity, coupled with periods of nagging doubt tinged with a flicker of terror. It was curiosity that started it all and terror that concluded it.

In 1934, following a lengthy series of discoveries in physics related to the phenomenon of radioactivity, Fermi discovered that interesting things happen when substances are bombarded with 'slow neutrons' – subatomic particles emitted by radioactive beryllium, and passed through paraffin to slow them down. Slow neutrons, Fermi discovered, were just what you needed to persuade other elements to emit their own radioactive particles. That looked interesting, so he squirted streams of slow neutrons at everything he could think of, and eventually he tried the then obscure element

* Or at least, less radioactive. We can but hope.

uranium, up until then mostly used as a source of yellow pigment. By something apparently like alchemy, the uranium turned into something new when the slow neutrons cannoned into it – but Fermi couldn't work out *what*.

Four years later three Germans – Otto Hahn, Lise Meitner, and Fritz Strassmann – repeated Fermi's experiments, and being better chemists, they worked out what had happened to the uranium. Mysteriously, it had turned into barium, krypton, and a small quantity of other stuff. Meitner realized that this process of 'nuclear fission' produced energy, by a remarkable method. Everyone knew that chemistry could turn matter into other kinds of matter, but now some of the matter in uranium was being transformed into *energy*, something that nobody had seen before. It so happened that Albert Einstein had already predicted this possibility on theoretical grounds, with his famous formula – an equation which the orangutan Librarian of Unseen University* would render as 'Ook'.†
Einstein's formula tells us that the amount of energy 'contained' in a given amount of matter is equal to the mass of that matter, multiplied by the speed of light and then multiplied by the speed of light again. As Einstein had immediately noticed, light is so fast it doesn't even appear to move, so its speed is decidedly big ... and the speed multiplied by itself is *huge*. In other words: you can get an awful lot of energy from a tiny bit of matter, if only you can find a way to do it. Now Meitner had worked out the trick.

A single equation may or may not halve your book sales, but it can change the world completely.

---

* He was the victim of a magical accident, which he rather enjoyed. But you *know* this.

† They say that every formula halves the sales of a popular science book. This is rubbish – if it was true, then *The Emperor's New Mind* by Roger Penrose would have sold one-eighth of a copy, whereas its actual sales were in the hundreds of thousands. However, just in case there is some truth to the myth, we have adopted this way of describing the formula to double our potential sales. You all know which formula we mean. You can find it written out in symbols on page 118 of Stephen Hawking's *A Brief History of Time* – so if the myth is right, he could have sold twice as many copies, which is a mindboggling thought.

Hahn, Meitner and Strassmann published their discovery in the British scientific journal *Nature* in January 1939. Nine months later Britain was at war, a war which would be ended by a military application of their discovery. It is ironic that the greatest scientific secret of World War II was given away just before the war began, and it shows how unaware politicians then were of the potential – be it for good or bad – of Big Science. Fermi saw the implications of the *Nature* article immediately, and he called in another top-ranking physicist, Niels Bohr, who came up with a novel twist: the chain reaction. If a particular, rare form of uranium, called uranium-235, was bombarded with slow neutrons, then not only would it split into other elements and release energy – it would also release more neutrons. Which, in turn, would bombard more uranium-235 … The reaction would become self-sustaining, and the potential release of energy would be *gigantic*.

Would it work? Could you get 'something for nothing' in this way? Finding out was never going to be easy, because uranium-235 is mixed up with ordinary uranium (uranium-238), and getting it out is like looking for a needle in a haystack when the needle is made of straw.

There were other worries *too* … in particular, might the experiment be too successful, setting off a chain reaction that not only spread through the experiment's supply of uranium-235, but through everything else on Earth as well? Might the atmosphere catch fire? Calculations suggested: probably not. Besides, the big worry was that if the Allies didn't get nuclear fission working soon then the Germans would beat them to it. Given the choice between our blowing up the world and *the enemy* blowing up the world, it was obvious what to do.

That is, on reflection, not a happy sentence.

Loko is remarkably similar to Oklo in southeastern Gabon, where there are deposits of uranium. In the 1970s, French scientists unearthed evidence that some of that uranium had either been undergoing unusually intense nuclear reactions or was much, much older than the rest of the planet.

It *could* have been an archaeological relic of some ancient civilization whose technology had got as far as atomic power, but a duller if more plausible expanation is that Oklo was a 'natural reactor'. For some accidental reason, that particular patch of uranium was richer than usual in uranium–235, and a spontaneous chain reaction ran for hundreds of thousands of years. Nature got there well ahead of Science, and without the squash court.

Unless, of course, it *was* an archaeological relic of some ancient civilization.

Until late in 1998, the natural reactor at Oklo was also the best evidence we could find to show that one of the biggest 'what if?' questions in science had an uninteresting answer. This question was 'What if the natural constants *aren't*?'

Our scientific theories are underpinned by a variety of numbers, the 'fundamental constants'. Examples include the speed of light, Planck's constant (basic to quantum mechanics), the gravitational constant (basic to gravitational theory), the charge on an electron, and so on. All of the accepted theories assume that these numbers have always been the same, right from the very first moment when the universe burst into being. Our calculations about that early universe *rely* on those numbers having been the same; if they used to be different, we don't know what numbers to put into the calculations. It's like trying to do your income tax when nobody will tell you the tax rates. From time to time maverick scientists advance the odd 'what if?' theory, in which they try out the possibility that one or more of the fundamental constants isn't. The physicist Lee Smolin has even come up with a theory of evolving universes, which bud off baby universes with different fundamental constants. According to this theory, our own universe is particularly good at producing such babies, and is also particularly suited to the development of life. The conjunction of these two features, he argues, is not accidental (the wizards at UU, incidentally, would be quite at home with ideas like this – in fact, sufficiently advanced physics is indistinguishable from magic).

Oklo tells us that the fundamental constants have not changed during the last two billion years – about half the age of the Earth and ten per cent of that of the universe. The key to the argument is a particular combination of fundamental constants, known as the 'fine structure constant'.* Its value is very close to 1/137 (and a lot of ink was devoted to explanations of that whole number 137, at least until more accurate measurements put its value at 137.036). The advantage of the fine structure constant is that its value does not depend on the chosen units of measurement – unlike say, the speed of light, which gives a different number if you express it in miles per second or kilometres per second. The Russian physicist Alexander Shlyakhter analysed the different chemicals in the Oklo reactor's 'nuclear waste', and worked out what the value of the fine structure constant must have been two billion years ago when the reactor was running. The result was the same as today's value to within a few parts in ten million.

In late 1998, though, a team of astronomers led by John Webb made a very accurate study of the light emitted by extremely distant, but very bright, bodies called quasars. They found subtle changes in certain features of that light, called spectral lines, which are related to the vibrations of various types of atom. In effect, what they seem to have discovered is that many billion years ago – much further back than the Oklo reactor – atoms didn't vibrate at quite the same rate as they do today. In very old gas clouds from the early universe, the fine structure constant differs from today's value by one part in 50,000. That's a *huge* amount by the standards of this particular area of physics. As far as anyone can tell, this unexpected result is not due to experimental error. A theory suggested in 1994 by Thibault Damour and Alexander Polyakov does indicate a possible variation in the fine structure constant, but only one-ten thousandth as large as that found by Webb's team.

In 2001 another team led by John Webb analysed the absorption of light from distant quasars by clouds of interstellar gas, and announced that the fine structure constant has increased by one hundred thousandth since the Big Bang. This result, if true, implies

* The fine structure constant is defined to be the square of the charge of an electron, divided by 2 times Planck's constant times the speed of light times the permittivity of the vacuum (as a handy lie, the last term might be thought as 'the way it reacts to an electric charge'). Thank you.

that the gravitational, weak nuclear, and strong nuclear forces have all changed over time. The work is being checked for possible systematic errors, which might produce the same observations.

It's all a bit of a puzzle, and most theorists sensibly prefer to hedge their bets and wait for further research. But it could be a straw in the wind: perhaps we will soon have to accept that the laws of physics were subtly different in the distant reaches of time and space. Not turtle-shaped, perhaps, but … different.

THREE
I KNOW
MY WIZARDS

 IT DID NOT TAKE LONG for the faculty to put its collective finger on the philosophical nub of the problem, *vis-à-vis* the complete destruction of everything.

'If no one will know if it happens, then in a very real sense it wouldn't have happened,' said the Lecturer in Recent Runes. His bedroom was on one of the colder sides of the university.

'Certainly we wouldn't get the blame,' said the Dean, 'even if it did.'

'As a matter of fact,' Ponder went on, emboldened by the wizards' relaxed approach, 'there is some theoretical evidence to suggest that it could not possibly happen, due to the non-temporal nature of the thaumic component.'

'Say again?' said Ridcully.

'A malfunction would not result in an *explosion* exactly, sir,' said Ponder. 'Nor, as far as I can work out, would it result in things ceasing to exist from the present onwards. They would cease to have existed at all, because of the multidirectional collapse of the thaumic field. But since we *are* here, sir, we must be living in a universe where things did not go wrong.'

'Ah, I know this one,' said Ridcully. 'This is because of quantum, isn't it? And there's some usses in some universe next door where it *did* go wrong, and the poor devils got blown up?'

'Yes, sir. Or, rather, no. They didn't get blown up because the device the other Ponder Stibbonses would have built would have gone wrong, and so ... he didn't exist not to build it. That's the theory, anyway.'

'I'm glad that's sorted out, then,' said the Senior Wrangler briskly. 'We're here because we're here. And since we're here, we

28

might as well be warm.'

'Then we seem to be in agreement,' said Ridcully. 'Mr Stibbons, you may start this infernal engine.' He nodded towards the red lever on the plinth.

'I was rather assuming you would do the honours, Archchancellor,' said Ponder, bowing. 'All you need to do is pull the lever. That will, ahem, release the interlock, allowing the flux to enter the exchanger, where a simple octiron reaction will turn the magic into heat and warm up the water in the boiler.'

'So it really *is* just a big kettle?' said the Dean.

'In a manner of speaking, yes,' said Ponder, trying to keep his face straight.

Ridcully grasped the lever.

'Perhaps you would care to say a few words, sir?' said Ponder.

'Yes.' Ridcully looked thoughtful for a moment, and then brightened up. 'Let's get this over quickly, and have lunch.'

There was a smattering of applause. He pulled the lever. The hand on a dial on the wall moved off zero.

'Well, we're not blown up after all,' said the Senior Wrangler. 'What are the numbers on the wall for, Stibbons?'

'Oh, er . . . they're . . . they're to tell you what number it's got to,' said Ponder.

'Oh. I see.' The Senior Wrangler grasped the lapels of his robe. 'Duck with green peas today, gentlemen, I believe,' he said, in a far more interested tone of voice. 'Well done, Mr Stibbons.'

The wizards ambled off in the apparently slow yet deceptively fast way of wizards heading towards food.

Ponder breathed a sigh of relief, which turned into a gulp when he realized that the Archchancellor had not, in fact, left but was inspecting the engine quite closely.

'Er . . . is there anything else I can tell you, sir?' he said, hurriedly.

'When did you *really* start it, Mister Stibbons?'

'Sir?'

'Every single word in the sentence was quite short and easy to understand. Was there something wrong about the way I assembled them?'

'I ... we ... it was started just after breakfast, sir,' said Ponder meekly. 'The needle on the dial was just turned by Mr Turnipseed by means of a string, sir.'

'Did it blow up at all when you started it up?'

'No sir! You'd ... well, you'd have *known*, sir!'

'I thought you said back there that we wouldn't have known, Stibbons.'

'Well, no, I mean –'

'I know you, Stibbons,' said Ridcully. 'And you would *never* test something out publicly before trying it to see if it worked. No one wants egg all over their face, do they?'

Ponder reflected that egg on the face is only of minor concern when the face is part of a cloud of particles expanding outwards at an appreciable fraction of the speed of dark.*

Ridcully slammed his hand against the black panels of the engine, causing Ponder visibly to leave the ground.

'Warm already,' he said. 'You all right up there, Bursar?'

The Bursar nodded happily.

'Good man. Well done, Mister Stibbons. Let's have lunch.'

After a while, when the footsteps had died away, it dawned on the Bursar that he was, as it were, holding the short end of the string.

The Bursar was not, as many thought, insane. On the contrary, he was a man with both feet firmly on the ground, the only difficulty being that the ground in question was on some other planet, the one with the fluffy pink clouds and the happy little bunnies. He did not mind because he much preferred it to the real one, where people shouted too much, and he spent as little time there as possible. Unfortunately this had to include mealtimes. The meal service on Planet Nice was unreliable.

Smiling his faint little smile, he put down his axe and ambled off. After all, he reasoned, the point was that the wretched thing stayed out of the ... whatever it was, and it could certainly do a simple job like that without his watching it.

* As yet unmeasured, but believed to be faster than light owing to its ability to move so quickly out of light's way.

Unfortunately Mr Stibbons was too worried to be very obser-vant, and none of the other wizards bothered much about the fact that everything which stood between them and thaumic devastation was blowing bubbles into his glass of milk.

FOUR
SCIENCE
AND MAGIC

 IF WE WANTED TO, we could comment on several features of Ponder Stibbons's experiment, describing the associated science. For example, there is a hint of the 'many worlds' interpretation of quantum mechanics, in which billions of universes branch off from ours every time a decision might go more than one way. And there is the unofficial standard procedure of public opening ceremonies, in which A Royal Personage or The President pulls a big lever or pushes a big button to 'start' some vast monument to technology – which has been running for days behind the scenes. When Queen Elizabeth II opened Calder Hall, the first British nuclear power station, this is just what went on – big meter and all.

However, it's a bit early for Quantum, and most of us have forgotten Calder Hall completely. In any case there's a more urgent matter to dispose of. This is the relation between science and magic. Let's start with science.

Human interest in the nature of the universe, and our place within it, goes back a long, long way. Early humanoids living on the African savannahs, for instance, can hardly have failed to notice that at night the sky was full of bright spots of light. At what stage in their evolution they first began to wonder what those lights were is a mystery, but by the time they had evolved enough intelligence to poke sticks into edible animals and to use fire, it is unlikely that they could stare at the night sky without wondering what the devil it was *for* (and, given humanity's traditional obsessions, whether it involved sex in some way). The Moon was certainly impressive – it was big, bright, and *changed shape*.

Creatures lower on the evolutionary ladder were certainly aware of the Moon. Take the turtle, for instance – about as Discwordly a

32

beast as you can get. When today's turtles crawl up the beach to lay their eggs and bury them in the sand, they somehow choose their timing so that when the eggs hatch, the baby turtles can scramble towards the sea by aiming at the Moon. We know this because the lights of modern buildings confuse them. This behaviour is remarkable, and it's not at all satisfactory to put it down to 'instinct' and pretend that's an answer. What *is* instinct? How does it work? How did it arise? A scientist wants plausible answers to such questions, not just an excuse to stop thinking about them. Presumably the baby turtles' moonseeking tendencies, and their mothers' uncanny sense of timing, evolved together. Turtles that just happened, by accident, to lay their eggs at just the right time for them to hatch when the Moon would be to seawards of their burial site, *and* whose babies just happened to head towards the bright lights, got more of the next generation back to sea than those that didn't. All that was needed to establish these tendencies as a universal feature of turtlehood was some way to pass them on to the next generation, which is where genes come in. Those turtles that stumbled on a workable navigational strategy, and could pass that strategy on to their offspring by way of their genes, did better than the others. And so they prospered, and outcompeted the others, so that soon the only turtles around were the ones that could navigate by the Moon.

Does Great A'Tuin, the turtle that holds up the elephants that hold up the Disc, swim through the depths of space in search of a distant light? Perhaps. According to *The Light Fantastic*, 'Philosophers have debated for years about where Great A'Tuin might be going, and have often said how worried they are that they might never find out. They're due to find out in about two months. And then they're *really* going to worry ...' For, like its earthbound counterpart, Great A'Tuin is in reproductive mode, in this case going to its own hatching ground to watch the emergence. That story ends with it swimming off into the cool depths of space, orbited by eight baby turtles (who appear to have gone off on their own, and perhaps even now support very small Discworlds) ...

The interesting thing about the terrestrial turtlish trickery is that at no stage is it necessary for the animals to be conscious that their timing is geared to the Moon's motion, or even that the Moon exists. However, the trick won't work unless the baby turtles *notice* the Moon, so we deduce that they did. But we can't deduce the existence of some turtle astronomer who wondered about the Moon's puzzling changes of shape.

When a particular bunch of social-climbing monkeys arrived on the scene, however, they began to ask such questions. The better the monkeys got at *answering* those questions, the more baffling the universe became; knowledge increases ignorance. The message they got was: *Up There is very different from Down Here*.

They didn't know that Down Here was a pretty good place for creatures like them to live. There was air to breathe, animals and plants to eat, water to drink, land to stand on, and caves to get out of the rain and the lions. They *did* know that it was changeable, chaotic, unpredictable …

They didn't *know* that Up There – the rest of the universe – isn't like that. Most of it is empty space, a vacuum. You can't breathe vacuum. Most of what isn't vacuum is huge balls of overheated plasma. You can't stand on a ball of flame. And most of what isn't vacuum and isn't burning is lifeless rock. You can't eat rock.* They were going to learn this later on. What they *did* know was that Up There was, in human timescales, calm, ordered, regular. And predictable, too – you could set your stone circle by it.

All this gave rise to a general feeling that Up There was different from Down Here for a *reason*. Down Here was clearly designed for *us*. Equally clearly, Up There wasn't. Therefore it must be designed for *somebody else*. And the new humanity was already speculating about some suitable tenants, and had been ever since they'd hidden in the caves from the thunder. The gods! They were Up There, looking Down! And they were clearly in charge, because humanity certainly wasn't. As a bonus, that explained all of the things Down Here that were a lot more complicated than anything visible Up There, like thunderstorms and earthquakes and bees. Those were under the control of the gods.

* Actually you *can* eat salt. But nobody outside Discworld goes to a restaurant to order a basalt balti.

It was a neat package. It made us feel important. It certainly made the priests important. And since priests were the sort of people who could have your tongue torn out or banish you into Lion Country for disagreeing with them, it rapidly became an enormously popular theory, if only because those who had other ones either couldn't speak or were up a tree somewhere.

And yet ... every so often some lunatic with no sense of self-preservation was born who found the whole story unsatisfying, and risked the wrath of the priesthood to say so. Such folk were already around by the time of the Babylonians, whose civilization flourished between and around the Tigris and Euphrates rivers from 4000 BC to 300 BC. The Babylonians – a term that covers a whole slew of semi-independent peoples living in separate cities such as Babylon, Ur, Nippur, Uruk, Lagash, and so on – certainly worshipped the gods like everyone else. One of their stories about gods is the basis of the Biblical tale of Noah and his ark, for instance. But they also took a keen interest in what those lights in the sky *did*. They knew that the Moon was round – a sphere rather than a flat disc. They probably knew that the Earth was round, too, because it cast a rounded shadow on the Moon during lunar eclipses. They knew that the year was about 365¼ days long. They even knew about the 'precession of the equinoxes', a cyclic variation that completes one cycle every 26,000 years. They made these discoveries by keeping careful records of how the Moon and the planets moved across the sky. Babylonian astronomical records from 500 BC survive to this day.

From such beginnings, an alternative explanation of the universe came into being. It didn't involve gods, at least directly, so it didn't find much favour with the priestly class. Some of their descendants are still trying to stamp it out, even today. The traditional priesthoods (who then and now often included some very intelligent people) eventually worked out an accommodation with this godless way of thinking, but it's still not popular with postmodernists, creationists, tabloid astrologers and others who prefer the answers you can make up for yourself at home.

The current name for what has variously been called 'heresy' and 'natural philosophy' is, of course, 'science'.

Science has developed a very strange view of the universe. It thinks that the universe runs on *rules*. Rules that never get broken. Rules that leave little room for the whims of gods.

This emphasis on rules presents science with a daunting task. It has to explain how a lot of flaming gas and rocks Up There, obeying simple rules like 'big things attract small things, and while small things also attract big things they don't do it strongly enough so as you'd notice', can have any chance whatsoever of giving rise to Down Here. Down Here, rigid obedience to rules seems notably absent. One day you go out hunting and catch a dozen gazelles; next day a lion catches *you*. Down Here the most evident rule seems to be 'There are no rules', apart perhaps from the one that could be expressed scientifically as 'Excreta Occurs'. As the Harvard Law of Animal Behaviour puts it: 'Experimental animals, under carefully controlled laboratory conditions, do what they damned well please.' Not only animals: every golfer knows that something as simple as a hard, bouncy sphere with a pattern of tiny dots on it never does what it's supposed to do. And as for the weather ...

Science has now divided into two big areas: the life sciences, which tell us about living creatures, and the physical sciences, which tell us about everything else. Historically, 'divides' is definitely the word – the scientific styles of these two big divisions have about as much in common as chalk and cheese. Indeed, chalk is a rock and so clearly belongs to the geological sciences, whereas cheese, formed by bacterial action on the bodily fluids of cows, belongs to the biological sciences. Both divisions are definitely science, with the same emphasis on the role of experiments in testing theories, but their habitual thought patterns run along different lines.

At least, until now.

As the third millennium approaches, more and more aspects of science are straddling the disciplines. Chalk, for instance, is more than just a rock: it is the remains of shells and skeletons of millions of tiny ocean-living creatures. And making cheese relies on chemistry and sensor technology as much as it does on the biology of grass and cows.

The original reason for this major bifurcation in science was a strong perception that life and non-life are extremely different. Non-life is simple and follows mathematical rules; life is complex and follows no rules whatsoever. As we said, Down Here looks very different from Up There.

However, the more we pursue the implications of mathematical rules, the more flexible a rule-based universe begins to seem. Conversely, the more we understand biology, the more important its physical aspects become – because life isn't a special *kind* of matter, so it too must obey the rules of physics. What looked like a vast, unbridgeable gulf between the life sciences and the physical sciences is shrinking so fast that it's turning out to be little more than a thin line scratched in the sand of the scientific desert.

If we are to step across that line, though, we need to revise the way we think. It's all too easy to fall back on old – and inappropriate – habits. To illustrate the point, and to set up a running theme for this book, let's see what the engineering problems of getting to the Moon tell us about how living creatures work.

The main obstacle to getting a human being on to the Moon is not distance, but gravity. You could *walk* to the Moon in about thirty years – given a path, air, and the usual appurtenances of the experienced traveller – were it not for the fact that it's uphill most of the way. It takes energy to lift a person from the surface of the planet to the neutral point where the Moon's pull cancels out the Earth's. Physics provides a definite lower limit for the energy you must expend – it's the difference between the 'potential energy' of a mass placed at the neutral point and the potential energy of the same mass placed on the ground. The Law of Conservation of Energy says that you can't do the job with less energy, however clever you are.

You can't beat physics.

This is what makes space exploration so expensive. It takes a lot of fuel to lift one person into space by rocket, and to make matters worse, you need more fuel to lift the *rocket* ... and more fuel to lift the fuel ... and ... At any rate, it seems that we're stuck at the bottom of the Earth's gravity well, and the ticket out *has* to cost a fortune.

Are we, though?

At various times, similar calculations have been applied to living creatures, with bizarre results. It has been 'proved' that kangaroos can't jump, bees can't fly, and birds can't get enough energy from their food to power their search for the food in the first place. It has even been 'proved' that life is impossible because living systems become more and more ordered, whereas physics implies that all systems become more and more disordered. The main message that biologists have derived from these exercises has been a deep scepticism about the relevance of physics to biology, and a comfortable feeling of superiority, because life is clearly much more interesting than physics.

The correct message is very different: be careful what tacit assumptions you make when you do that kind of calculation. Take that kangaroo, for instance. You can work out how much energy a kangaroo uses when it makes a jump, count how many jumps it makes in a day, and deduce a lower limit on its daily energy requirements. During a jump, the kangaroo leaves the ground, rises, and drops back down again, so the calculation is just like that for a space rocket. Do the sums, and you find that the kangaroo's daily energy requirement is about ten times as big as anything it can get from its food. Conclusion: kangaroos can't jump. Since they can't jump, they can't find food, so they're all dead.

Strangely, Australia is positively teeming with kangaroos, who fortunately cannot do physics.

What's the mistake? The calculation models a kangaroo as if it were a sack of potatoes. Instead of a thousand kangaroo leaps per day (say), it works out the energy required to lift a sack of potatoes off the ground and drop it back down, 1000 times. But if you look at a slow-motion film of a kangaroo bounding across the Australian outback, it doesn't look like a sack of potatoes. A kangaroo *bounces*, lolloping along like a huge rubber spring. As its legs go up, its head and tail go down, storing energy in its muscles. Then, as the feet hit the ground, that energy is released to trigger the next leap. Because most of the energy is borrowed and paid back, the energy cost per leap is tiny.

Here's an association test for you. 'Sack of potatoes' is to 'kangaroo' as 'rocket' is to – *what*? One possible answer is a space elevator. In the October 1945 issue of *Wireless World* the science-fiction writer Arthur C. Clarke invented the concept of a geostationary orbit, now the basis of virtually all communications satellites. At a particular height – about 22,000 miles (35,000 km) above the ground – a satellite will go round the Earth exactly in synchrony with the Earth's rotation. So from the ground it will look as though the satellite isn't moving. This is useful for communications: you can point your satellite dish in a fixed direction and always get coherent, intelligent signals or, failing that, MTV.

Nearly thirty years later Clarke popularized a concept with far greater potential for technological change. Put up a satellite in geostationary orbit and drop a long cable down to the ground. It has to be an amazingly strong cable: we don't yet have the technology but 'carbon nanotubes' now being created in the laboratory come close. If you get the engineering right, you can build an elevator 22,000 miles high. The cost would be enormous, but you could then haul stuff into space just by pulling on the cable from above.

Ah, but you can't beat physics. The energy required would be exactly the same as if you used a rocket.

Of course. Just as the energy required to lift a kangaroo is exactly the same as that required to lift a sack of potatoes.

The trick is to find a way to borrow energy and pay it back. The point is that once the space elevator is in place, after a while there's just as much stuff coming down it as there is going up. Indeed, if you're mining the Moon or the asteroids for metals, there will soon be *more* stuff coming down than goes up. The materials going down provide the lifting energy for those going up. Unlike a rocket, which gets used up every time you fire it, a space elevator is self-sustaining.

Life is like a space elevator. What life self-sustains is not energy, but organization. Once you have a system that is so highly organized that it can reliably make copies of itself, that degree of organization is no longer 'expensive'. The initial investment may have been huge, as for a space elevator, but once the investment has been made, everything else is free.

If you want to understand biology, it is the physics of space elevators that you need, not the physics of rockets.

How can Discworld's magic illuminate Roundworld's science? Just as the gulf between the physical and biological sciences is turning out to be far narrower than we used to think, so the gulf between science and magic is also becoming smaller. The more advanced our technologies become, the less possible it is for the everyday user to have any idea of how they work. As a result, they look more and more like magic. As Clarke realized, this tendency is inevitable; Gregory Benford went further and declared it desirable.

Technology works because whoever built it in the first place figured out enough of the rules of the universe to make the technology do what was required of it. You don't need to get the rules *right* to do this, just right *enough* – space rockets work fine even though their orbits are computed using Newton's stab at the rules of gravity, which aren't as accurate as Einstein's. But what you can accomplish is severely constrained by what the universe will permit. With magic, in contrast, things work because people want them to. You still have to find the right spell, but what drives the development is human wishes (and, of course, the knowledge, skill and experience of the practitioner). This is one reason why science often seems inhuman, because it looks at how the universe drives *us*, rather than the other way round.

Magic, however, is only one aspect of Discworld. There's a lot of science on Discworld, too – or at least rational engineering. Balls get thrown and caught, the biology of the river Ankh resembles that of a typical terrestrial swamp or sewage farm, and light goes in more or less straight lines. Very slowly, though. As we read in *The Light Fantastic:* 'Another Disc day dawned, but very gradually, and this is why. When light encounters a strong magical field it loses all sense of urgency. It slows right down. And on the Discworld the magic was embarrassingly strong, which meant that the soft yellow light of dawn flowed over the sleeping landscape like the caress of a gentle lover or, as some, would have it, like golden syrup.' The same quote tells us that as well as rational engineering there's a lot of magic in Discworld: overt magic which slows light down; magic that allows the sun to orbit the world provided that occasionally one of the elephants lifts its leg to let the sun pass. The sun is small, nearby, and travels faster than its own light. This appears to cause no major problems.

There is magic in our world, too, but of a different, less obvious kind. It happens around everybody all the time, in all those little causalities which we don't understand but just accept. When we turn the switch and the light comes on. When we get into the car and start the engine. When we do all those improbable and ridiculous things that, thanks to biological causality, make babies. Certainly many people understand, often to quite a detailed degree, what is going on in particular areas – but sooner or later we all reach our Magical Event Horizon. Clarke's Law states that any sufficiently advanced technology looks like magic. 'Advanced' here is usually taken to mean 'shown to us by advanced aliens or people from the future', like television shown to Neanderthals. But we should realize that television is magic to nearly everyone that uses it *now* – to those behind the camera as well as to those sitting on the couch in front of the moving picture in the funny box. At some point in the process, in the words of cartoonist S. Harris, 'a miracle occurs'.

Science takes on the aura of magic because the design of a civilization proceeds by a type of narrative imperative – it makes a coherent *story*. In about 1970, Jack gave a lecture to a school audience on 'The Possibility of Life on Other Planets'*. He talked about evolution, what planets were made of – all the things that you'd expect in such a lecture. The first question was from a girl of about 15, who asked 'You believe in evolution, don't you, sir?' The teacher went on about it not being a 'proper' question, but Jack answered it anyway, saying – rather pretentiously –'No, I don't *believe* in evolution, like people believe in God ... Science and technology are not advanced by people who believe, but by people who *don't know* but are doing their best to find out ... steam engine ... spinning jenny ... television ...' At that, she was on her feet again: 'No, that ain't how television was invented!' The teacher tried to calm the discussion by asking her to explain how she thought television was invented. 'My father works for Fisher Ludlow making

* Now turned into a book: *Evolving the Alien* by Jack Cohen and Ian Stewart.

pressed steel for car bodies. He gets paid and he gives some of the money to the government to give him things. So he tells the government he wants to watch television, and they pay someone to invent television, and they do!'

It's very easy to make this mistake, because technology progresses by pursuing goals. We get the feeling that if we pour in enough resources, we can achieve anything. Not so. Pour in enough resources, and we can achieve *anything* that is within reach of current know-how, or possibly just a bit beyond if we're lucky. But nobody tells us about the inventions that fail. Nobody tries to raise funding for a project that they know can't *possibly* work. No funding body will pay for research projects in which we have no idea where to start. We could pour as much money as we liked into developing antigravity or faster-than-light travel, and we'd get nowhere.

When you can take a machine to bits and see how it works, you get a clear feeling for the constraints within which it has to operate. In such cases, you're not going to confuse science and magic. The first cars required an extremely hands-on starting system – you stuck a big handle into the engine and literally 'turned it over'. Whatever the engine did when it started, you knew it wasn't magic. However, as technology develops it usually doesn't *remain* transparent to the user. As more people began to use cars, more and more of the obvious technology was replaced by symbols. You worked switches with labels to get things to happen. That's our version of the magic spell: you pull a knob called Cold Start and the engine does all the cold start things for itself. When Granny wants to drive she does not have to do much more than push the accelerator for 'Go'. Little imps do the rest, by magic.

This process is the core of the relation between science and magic in our own world. The universe into which we were born, and in which our species evolved, runs by rules – and science is our way of trying to work out what the rules are. But the universe that we are now constructing for ourselves is one that, to anyone other than a member of the design team and very possibly even to them, works by magic.

A special kind of magic is one of the many things that have made humans what they are. It's called education. It's how we pass on ideas from one generation to the next. If we were like computers, we'd be able to *copy* our minds into our children, so that they would grow up agreeing with every opinion that we hold dear. Well, actually they wouldn't, though they might start out that way. There is an aspect of education that we want to draw to your attention. We call it 'lies-to-children'. We're aware that some readers may object to the word 'lie' – it got Ian and Jack into terrible trouble with some literally-minded Swedes at a scientific conference who took it all terribly seriously and spent several days protesting that 'It's not a *lie*!' It is. It is for the best possible reasons, but it is still a lie. A lie-to-children is a statement that is false, but which nevertheless leads the child's mind towards a more accurate explanation, one that the child will only be able to appreciate if it has been primed with the lie.

The early stages of education *have* to include a lot of lies-to-children, because early explanations have to be simple. However, we live in a complex world, and lies-to-children must eventually be replaced by more complex stories if they are not to become delayed-action genuine lies. Unfortunately, what most of us know about science consists of vaguely remembered lies-to-children. For example, the rainbow. We all remember being told at school that glass and water split light into its constituent colours – there's even a nice experiment where you can *see* them – and we were told that this is how rainbows form, from light passing through raindrops. When we were children, it never occurred to us that while this explains the colours of the rainbow it doesn't explain its shape. Neither does it explain how the light from the many different raindrops in a thundershower somehow combines to create a bright arc. Why doesn't it all smudge out? This is not the place to tell you about the elegant geometry of the rainbow – but you can see why 'lie' is not such a strong word after all. The school explanation diverts our attention from the real marvel of the rainbow, the cooperative effects of all the raindrops, by trying to pretend that once you've explained the colours, that's it.

Other examples of lies-to-children are the idea that the Earth's magnetic field is like a huge bar magnet with N and S marked on it, the picture of an atom as a miniature solar system, the idea that a living amoeba is a billion-year-old 'primitive' organism, the image of DNA as the blueprint for a living creature, and the connection between relativity and Einstein's hairstyle (it's the sort of crazy idea that only people with hair like *that* can come up with). Quantum mechanics lacks a public 'icon' of this kind – it doesn't tell a simple story that a non-specialist can grab and hang on to – so we feel uncomfortable about it.

When you live in a complex world, you have to simplify it in order to understand it. Indeed, that's what 'understand' means. At different stages of education, different levels of simplification are appropriate. Liar-to-children is an honourable and vital profession, otherwise known as 'teacher'. But what teaching does *not* do – although many politicians think it does, which is one of the problems – is erect a timeless edifice of 'facts'.* Every so often, you have to unlearn what you thought you already knew, and replace it by something more subtle. This *process* is what science is all about, and it never stops. It means that you shouldn't take everything *we* say as gospel, either, for we belong to another, equally honourable profession: Liar-to-readers.

On Discworld, one of Ponder Stibbons's lies-to-wizards is about to come seriously unstuck.

---

* As humans, we have invented lots of useful kinds of lie. As well as lies-to-children ('as much as they can understand') there are lies-to-bosses ('as much as they need to know') lies-to-patients ('they won't worry about what they don't know') and, for all sorts of reasons, lies-to-*ourselves*. Lies-to-children is simply a prevalent and necessary kind of lie. Universities are very familiar with bright, qualified school-leavers who arrive and then go into shock on finding that biology or physics isn't quite what they've been taught so far. 'Yes, but you needed to understand *that*,' they are told, 'so that *now* we can tell you why it isn't exactly *true*.' Discworld teachers know this, and use it to demonstrate why universites are truly storehouses of knowledge: students arrive from school confident that they know very nearly everything, and they leave years later certain that they know practically nothing. Where did the knowledge go in the meantime? Into the university, of course, where it is carefully dried and stored.

FIVE

THE ROUNDWORLD PROJECT

ARCHCHANCELLOR RIDCULLY AWOKE FROM AN AFTER-NOON NAP in which he had been crawling through a baking desert under a flamethrower sky, and found that this was more or less true.

Superheated steam whistled from the joints of the radiator in the corner. Ridcully walked over through the stifling air and touched it gently.

'Ouch! Damnation!'

Sucking his right hand and using his left hand to unwrap the scarf from his neck, he strode out into the corridor and what looked like Hell with the heat turned up. Steam rolled along the corridors, and from somewhere overhead came the once-heard-never-forgotten *thwack* of a high-energy magical discharge. Violet light filled the windows for a moment.

'Will someone tell me what the *heck* is going on?' Ridcully demanded of the air in general.

Something like an iceberg loomed out of the steam. It was the Dean.

'I would like to make it *absolutely* clear, Archchancellor, that this is *nothing* to do with me!'

Ridcully wiped away the sweat that was beginning to trickle down his forehead.

'Why are you standin' there in just your drawers, Dean?'

'I – well, my room is absolutely *boiling* hot –'

'I demand you put *something* on, man, you look thoroughly unhygienic!'

There was another crack of discharged magic. Sparks flew off the end of Ridcully's fingers.

'I *felt* that one!' he said, running back into his room.

Beyond the window, on the other side of the gardens, the air wavered over the High Energy Magic building. As the Archchancellor watched, the two huge bronze globes on its roof became covered in crawling, zig-zagging purple lines –

He hit the floor rolling, as wizards are wont to do, just before the shock of the discharge blew the windows in.

Melted snow was pouring off the rooftops. Every icicle was a streaming finger of water.

A large door bumped and scraped its way across the steaming lawns.

'For goodness' sake, Dean, handle your end, can't you?'

The door skidded a little further.

'It's no good, Ridcully, it's solid oak!'

'And I'm very glad of it!'

Behind Ridcully and the Dean, who were inching the door forward largely by arguing with each other, the rest of the faculty crept forward.

The bronze globes were humming now, in the rapidly decreasing intervals between discharges. They had been installed, to general scoffing, as a crude method of releasing the occasional erratic build-up of disorganized magic in the building. Now they were outlined in unhealthy-looking light.

'And we know what that means, don't we, Mister Stibbons?' said Ridcully, as they reached the entrance to the High Energy Magic building.

'The fabric of reality being unravelled and leaving us prey to creatures from the Dungeon Dimensions, sir?' mumbled Stibbons, who was trailing behind.

'That's *right*, Mister Stibbons! And we don't want that, do we, Stibbons?'

'No, sir.'

'No, sir! We don't, sir!' Ridcully roared. 'It'll be tentacles all over the place again. And none of us wants tentacles all over the place, do we?'

'No, sir.'

'No, sir! So switch the damned thing *off*, sir!'

'But it'd be certain death to go into –' Ponder stopped, swallowed and restarted. 'In fact it would be *uncertain* death to go into the squash court at the moment, Archchancellor. There must be million of thaums of random magic in there! Anything could happen!'

Inside the HEM the ceiling was vibrating. The whole building seemed to be dancing.

'They certainly knew how to build, didn't they, when they built the old squash court,' said the Lecturer in Recent Runes, in an admiring tone of voice. 'Of course, it *was built* to contain large amounts of magic ...'

'Even if we *could* switch it off, I don't think that'd be such a good idea,' said Ponder.

'Sounds a lot better than what's happening now,' said the Dean.

'But is falling through the air better than hitting the ground?' said Ponder.

Ridcully sucked in his breath between his teeth.

'That's a point,' he said. 'Could be something of an implosion, I suppose. You can't just *stop* something like this. Something bad would happen.'

'The end of the world?' quavered the Senior Wrangler.

'Probably just this part of it,' said Ponder.

'Are we talking here about a sort of huge valley about twenty miles across with mountains all round it?' said Ridcully, staring at the ceiling. Cracks were zig-zagging across it.

'Yes, sir, I'm wondering if whoever tried this at Loko actually *did* manage to switch it off ...'

The walls groaned. There was a rattling noise behind Ponder. He recognized it, even above the din. It was the sound of HEX resetting its writing device. Ponder always thought of it as a kind of mechanical throat-clearing.

The pen jerked in its complex network of threads and springs, and then wrote:

+++ This May Be Time For The Roundworld Project +++

'What are you talking about, man?' snapped Ridcully, who'd never quite understood what HEX was.

'Oh, *that*? That's been around for *ages*,' said the Dean. 'No one's ever taken it seriously. It's just a thought experiment. You couldn't do it. It's completely absurd. It needs far too much magic.'

'Well, we've *got* far too much magic,' said Ridcully. 'Right now we need to use it up.'

There was a moment's silence. That is, the wizards were silent. Overhead, magic flared into the sky with a sound like roaring gas.

'Can't let it build up here,' Ridcully went on. 'What's the Roundworld project then?'

'It was, er ... there was once some suggestion that it might be possible to create a ... an area where the laws of magic don't apply,' said Ponder. 'We could use it to learn more about magic.'

'Magic's *everywhere*,' said Ridcully. 'It's part of what everywhere *is*.'

'Yes, sir,' said Ponder, watching the Archchancellor carefully.

The ceiling creaked.

'What use would it be, anyway?' said Ridcully, still thinking aloud.

'Well, sir, you could ask what use is a new-born child ...'

'No, that's not the sort of question I ask,' said Ridcully. 'And it's a highly suspicious one, too.'

The wizards ducked as the latest discharge crackled overhead. It was followed by a louder explosion.

'I think the balls have just exploded, sir,' said Ponder.

'All right, then, how long would the project take to set up?' said Ridcully.

'Months,' said the Dean firmly.

'We've got about ten seconds to the next discharge, sir,' said Ponder. 'Only ... now the balls have gone it will simply earth itself ...'

'Ah. Oh. Really? Well, then ...' Ridcully looked around at his fellow wizards as the wall began to shake again. 'It's been nice knowing you. Some of you. One or two of you, anyway ...'

The whine of increasing magic rose in pitch.

The Dean cleared his throat.

'I'd just like to say, Mustrum,' he began.

'Yes, old friend?'

'I'd just like to say ... I think I'd have made a much better Archchancellor than you.'

The whine stopped. The silence twanged. The wizards held

their breath.

Something went 'ping'.

A globe about a foot across hung in the air between the faculty. It looked like glass, or the sheen of a pearl without the pearl itself.

From the squash court next door there was, instead of the wild roar of disorganized thaums, the steady thrum-thrum of purpose.

'What the heck is *that*?' said Ridcully, as the wizards unfolded themselves.

HEX rattled. Ponder picked up the piece of paper.

'Well, according to this, it's the Roundworld Project,' he said. 'And it's absorbing all the energy from the thaumic pile.'

The Dean brushed some dust off his robe.

'Nonsense,' he said. 'Takes months. Anyway, how could that machine possibly know the spells?'

'Mr Turnipseed did copy in a lot of the grimoires last year,' said Ponder. 'It's vital that HEX knows basic spell structure, you see …'

The Senior Wrangler peered irritably at the sphere.

'Is this all it is?' he said. 'Doesn't seem much for all that effort.'

There was a frightening moment as the Dean walked up to the sphere and his nose, enormously magnified, appeared in it.

'Old Archchancellor Bewdley devised it,' he said. 'Everyone said it was impossible …'

'Mr Stibbons?' said Ridcully.

'Yes, sir?'

'Are we in danger of blowing up at the moment?'

'I don't think so, sir. The … project is sucking up everything.'

'Shouldn't it be glowing, then? Or something? What's in there?'

HEX wrote: +++ Nothing +++

'All that magic's going into empty space?'

+++ Empty Space Is Not Nothing, Archchancellor. There Is Not Even Empty Space Inside The Project. There Is No Time For It To Be Empty In +++

'What's it got in it, then?'

+++ I Am Checking +++, HEX wrote patiently.

'Look, I can stick my hand right in it,' said the Dean.

The wizards watched in horror. The Dean's fingers were visible, darkly, within the sphere, outlined in thousands of tiny sparkling lights.

'That was a really very foolish thing you just did,' said Ridcully. 'How did you know it wasn't dangerous?'

'I didn't,' said the Dean cheerfully. 'It feels … cool. And rather chilly. Prickly, in a funny sort of way.'

HEX rattled. Ponder walked back and looked down at the paper.

'It almost feels *sticky* when I move my fingers,' said the Dean.

'Er … Dean?' said Ponder, stepping back carefully. 'I think it would be a really good idea if you pulled your hand out very, very carefully and really very soon.'

'That's odd, it's beginning to tingle –'

'Right now, Dean! Right now!'

For once, the urgency in Ponder's voice got through the Dean's cosmic self-confidence. He turned to argue with Ponder Stibbons just a moment before a white spark appeared in the centre of the sphere and began to expand rapidly.

The sphere flickered.

'Anyone know what caused that?' said the Senior Wrangler, his face bathed in the growing light of the Project.

'I *think*,' said Ponder slowly, holding up HEX's write-out, 'it was Time and Space starting to happen.'

In HEX's careful writing, the words said: +++ In The Absence Of Duration And Dimension, There Must Be *Potentiality*. +++

And the wizards looked upon the universe that was growing within the little sphere and spake amongst themselves, saying, 'It's rather a small one, don't you think? Is it dinner time yet?'

Later on, the wizards wondered if the new universe might have been different if the Dean had waggled his fingers in a different way. Perhaps, within it, matter might have naturally formed itself into, say, garden furniture, or one giant nine-dimensional flower a trillion miles across. But Archchancellor Ridcully pointed out that this was not very useful thinking, because of the ancient principle of WYGIWYGAINGW.*

---

* 'What You Get Is What You're Given And It's No Good Whining.'

SIX

BEGINNINGS AND BECOMINGS

 POTENTIALITY IS THE KEY.

Our immediate task is to start from a lot of vacuum and a few rules, and convince you that they have enormous potentiality. Given enough time, they can lead to people, turtles, weather, the Internet – hold it. *Where did all that vacuum come from?* Either the universe has been around forever, or once there wasn't a universe and then there was. The second statement fits neatly with the human predilection for creation myths. It also appeals to today's scientists – possibly for the same reason. Lies-to-children run deep.

Isn't vacuum just … empty space? What was there before we had space? How do you make space? Out of vacuum? Isn't that a vicious circle? If in the past we didn't have space, how can there have been a 'there' for whatever it was to exist in? And if there wasn't anywhere for it to exist, how did it manage to make space? Maybe space was there all along … but why? And what about *time*? Space is easy compared to time. Space is just … somewhere to put matter. Matter is just … stuff. But time – time flows, time passes, time makes sense in the past and the future but not in the instantaneous, frozen present. What makes time flow? Could the flow of time be stopped? *What would happen if it did?*

There are little questions, there are medium-sized questions, and there are big questions. After which there are even bigger questions, huge questions, and questions so vast that it is hard to imagine what kind of response would count as an answer.

You can usually recognize the little questions: they look immensely complicated. Things like 'What is the molecular structure of the left-handed isomer of glucose?' As the questions get bigger, they become deceptively simpler: 'Why is the sky blue?' The *really* big questions are *so* simple that it seems astonishing that sci-

51

ence has absolutely no idea how to answer them: 'Why doesn't the universe run backwards instead?' or 'Why does red look like *that*?'

All this goes to show that it's a lot easier to ask a question than it is to answer it, and the more specialized your question is, the longer are the words that you must invent to state it. Moreover, the bigger a question is, the more people are interested in it. Hardly anybody cares about left-handed glucose, but nearly all of us wonder why red looks the way it does, and we vaguely wonder whether it looks the same to everybody else.

Out on the fringes of scientific thought are questions that are big enough to interest almost everybody, but small enough for there to be a chance of answering them reasonably accurately. They are questions like 'How did the universe begin?' and 'How will it end?' ('What happens in between?' is quite a different matter.) Let us acknowledge, right up front, that the current answers to such questions depend upon various questionable assumptions. Previous generations have been absolutely convinced that their scientific theories were well-nigh perfect, only for it to turn out that they had missed the point entirely. Why should it be any different for our generation? Beware of scientific fundamentalists who try to tell you everything is pretty much worked out, and only a few routine details are left to do. It is just when the majority of scientists believe such things that the next revolution in our world-view creeps into being, its feeble birth-squeaks all but drowned by the earsplitting roar of orthodoxy.

Let's take a look at the current view of how the universe began. One of the points we are going to make is that human beings have trouble with the concept of 'beginning'. And even more trouble, let it be said, with 'becoming'. Our minds evolved to carry out rather specific tasks like choosing a mate, killing bears with a sharp stick, and getting dinner without *becoming* it. We've been surprisingly good at adapting those modules to tasks for which they were never 'intended' – that is, tasks for which they were not *used* during their evolution, there being no conscious 'intention' – such as planning a route up the Matterhorn, carving images of sea-lions on polar bears' teeth,* and

---

* Not while these are still in the polar bear.

calculating the combustion point of a complex hydrocarbon molecule. Because of the way our mental modules evolved, we think of beginnings as being analogous to how a day begins, or how a hike across the desert begins; and we think of becomings in the same way that a polar bear's tooth becomes a carved amulet, or a live spider becomes dead when you squash it.

That is: beginnings start from somewhere (which is where whatever it is *begins*), and becomings turn Thing One into Thing Two by pushing it across a clearly defined boundary (the tooth was not carved, but now it is; the spider was not dead, but now it is). Unfortunately the universe doesn't work in such a simple-minded manner, so we have serious trouble thinking about how a universe can begin, or how an ovum and a sperm can become a living child.

Let us leave becomings for a moment, and think about beginnings. Thanks to our evolutionary prejudices, we tend to think of the beginning of the universe as being some special time, before which the universe did not exist and after which it did. Moreover, when the universe changed from not being there to being there, something must have *caused* that change – something that was around before the universe began, otherwise it wouldn't have been able to cause the universe to come into being. When you bear in mind that the beginning of the universe is also the beginning of space and the beginning of time, however, this point of view is distinctly problematic. How can there be a 'before' if time has not yet started? How can there be a cause for the universe starting up, without space for that cause to happen in – and time for it to happen?

Maybe there was something else in existence already … but now we have to decide how *that* got started, and the same difficulties arise. All right, let's go the whole hog: something – perhaps the universe itself, perhaps some precursor – was around forever. It didn't *have* a beginning, it just was, always.

Satisfied? Things that exist forever don't have to be explained, because they don't need a cause? Then what caused them to have been around forever?

It now becomes impossible not to mention the turtle joke. Stephen Hawking tells it at the start of *A Brief History of Time*, but it goes back a lot further. According to Hindu legend, the Earth rides on the back of four elephants, which ride on a turtle. But what supports the turtle? In Discworld, Great A'Tuin needs no support, swimming through the universe unperturbed by any thought about what holds it up. That's magic in action: world-carrying turtles are *like that*. But according to the old lady who espoused the Hindu cosmology, and was asked the same question by a learned astronomer, there is a different answer: 'It's turtles all the way down!' The image of an infinite pile of turtles is instantly ludicrous, and very few people find it a satisfying explanation. Indeed very few people find it a satisfying *kind* of explanation, if only because it doesn't explain what supports the infinite pile of turtles. However, most of us are quite content to explain the origins of time as 'it's always been there'. Seldom do we examine this statement closely enough to realize that what it really says is 'It's time all the way back.' Now replace 'time' by 'turtle' and 'back' by 'down' ... Each instant of time is 'supported', that is, a causal consequence of, the previous instant of time. Fine, but that doesn't explain why time exists. What caused that infinite expanse of time? What holds up the whole pile?

All of which puts us in a serious quandary. We have problems thinking of time as beginning *without* a precursor, because it's hard to see how the causality goes. But we have equally nasty problems thinking of time as beginning *with* a precursor, because then we hit the turtle-pile problem. We have similar problems with space: either it goes on forever, in which case it's 'space all the way out' and we need somewhere even bigger to put the whole thing, or it stops, in which case we wonder what's outside it.

The real point is that neither of these options is satisfactory, and the origins of space and time fit neither model. The universe is not like a village, which ends at a fence or an imaginary line on the ground, neither is it like the distant desert which seems to vanish into eternity but actually just gets too far away for us to see it clearly. Time is not like a human lifespan, which starts at birth and ends at death, nor is it like the extended lifespan found in many religions, where the human soul continues to live indefinitely after death – and the much rarer belief (held, for example, by Mormons) that some aspect of each person was somehow already alive in the indefinite past.

So how did the universe begin? 'Begin' is the wrong word. Nonetheless, there is good evidence that the age of the universe is about 15 billion years,* so nothing – not space, not time – existed before some instant of time roughly 15 billion years ago. See how our narrativium-powered semantics confuses us. This does *not* mean that if you went back 15 billion and one years, you would find nothing. It means that you cannot go back 15 billion and one years. That description makes no sense. It refers to *a time before time began*, which is logically incoherent, let alone physically so.

This is not a new point. Saint Augustine made it very clearly in his *Confessions* of about 400, when he said: '... if there was no time before heaven and earth were created, how can anyone ask what you [God] were doing "then"? If there was no time, there was no "then"'.

Some have gone to greater extremes. In *The End of Time*, Julian Barbour argues that time does not exist. In his view, the 'real' laws of physics specify a static universe of timeless 'nows'. He argues that time is not a physical dimension in the way that space is. The apparent flow of time is an illusion, created because the collection of nows that corresponds to a human being makes that human experience its world *as if* it were an ordered series of events, happening one after the other. Apparent gaps of 'duration' between successive events are merely blanks where no 'now' exists. *Thief of Time* explores some remarkably similar issues from the Discworld viewpoint. But for this book, we will take a conventional attitude towards time, and consider it to be a real part of physics.

Cosmologists are pretty sure that it all began like this. The universe came into being as a tiny speck of space and time. The amount

---

* This figure replaces the previously favoured value of about 20 billion years. Recently lots of scientists collectively decided it should be 15 billion instead. (For a while some stars seemed to be older than the universe, but the age of those stars has also been downsized.) In other circumstances they might well have settled for 20 billion. If this worries you, substitute the term 'a very long time'.

of space inside this tiny speck grew rapidly, and time began to elapse so that 'rapidly' actually had a meaning. Everything that there is, today, right out into the furthest depths of space, stems from that astonishing 'beginning'. Colloquially, the event is known as the Big Bang. The name reflects several features of the event – for example, that tiny speck of space/time was enormously hot, and grew in size exceedingly rapidly. It was like a huge explosion – but there was no stick of cosmic dynamite, sitting there in no-space with its non-material fuse burning away as some kind of pre-time pseudo-clock counted down the seconds to detonation. What exploded was – nothing. Space, time, and matter are the *products* of that explosion: they played no part in its cause. Indeed, in a very real sense, it *had* no cause.

The evidence in favour of the Big Bang is twofold. The first item is the discovery that the universe is expanding. The second is that 'echoes' from the Big Bang can still be detected today. The possibility that the universe might be getting bigger first appeared in mathematical solutions to equations formulated by Albert Einstein. Einstein viewed spacetime as being 'curved'. A body moving through curved spacetime deviates from its normal straight line path, much as a marble rolling on a curved surface does. This deviation can be interpreted as a 'force' – something that *pulls* the body away from that ideal straight line. Actually there is no pull: just a bend in spacetime, causing a bend in the body's path. But it looks as if there's a pull. Indeed this apparent pull is what Newton called 'gravity', back in the days when people thought it really did pull bodies together. Anyway, Einstein wrote down some equations for how such a bendy universe ought to behave. They were very difficult equations to solve, but after making some extremely strong assumptions – basically that at any instant of time space is a sphere – mathematical physicists worked out a few answers. And this tiny, very special list of solutions, the only ones their feeble methods could find, told them three things that the universe could do. It could stay the same size forever; it could collapse down to a single point; or it could start from a single point and grow in size without limit.

We now know that there are many other solutions to Einstein's equations, leading to all sorts of bizarre behaviour, but back in the days when today's paradigm was being set, these solutions were the only ones anybody knew. So they assumed that the universe must behave according to one or other of those three solutions. Science was subliminally prepared either for continuous creation (the universe is always the same) or for the Big Bang. The Big Crunch, in which the universe shrinks to an infinitely dense, infinitely hot point, lacked psychological appeal.

Enter Edwin Hubble, an American astronomer. Hubble was observing distant stars, and he made a curious discovery. The further away the stars were, the faster they were moving. He knew this for distinctly indirect – but scientifically impeccable – reasons. Stars emit light, and light has many different colours, including 'colours' that the human eye is unable to see, colours like infra-red, ultra-violet, radio, x-ray … Light is an electromagnetic wave, and there is one 'colour' for each possible wavelength of light – the distance from one electromagnetic peak to the next. For red light, this distance is 2.8 hundred thousandths of an inch (0.7 millionths of a metre).

Hubble noticed that something funny was happening to the light emitted by stars: the colours were shifting in the red direction. The further away a star was, the bigger the shift. He interpreted this 'red shift' as a sign that the stars are moving away from us, because there is a similar shift for sound, known as the 'Doppler effect', and it's caused by the source of the sound moving. So the further away the stars are, the faster they're travelling. This means that the stars aren't just moving away from *us* – they're moving away from each other, like a flock of birds dispersing in all directions.

The universe, said Hubble, is expanding.

Not expanding *into* anything, of course. It's just that the space inside the universe is *growing*.* That made the physicists' ears prick up, because it fitted exactly one of their three scenarios for changes in the size of the universe: stay the same, grow, collapse They

---

* Indeed, impeccable Discworld thinking is that no matter how big the universe grows, it's always the same size.

'knew' it had to be one of the three, but which? Now they knew that, too. If we accept that the universe is growing we can work out where it came from by running time backwards, and this time-reversed universe collapses back to a single point. Putting time the right way round again, it must all have grown *from* a single point – the Big Bang. By estimating the rate of expansion of the universe we can work out that the Big Bang happened about 15 billion years ago.

There is further evidence in the Big Bang's favour: it left 'echoes'. The Big Bang produces vast amounts of radiation, which spreads through the universe. Over billions of years, the remnants of the Big Bang's radiation smeared out into the 'cosmic background', a kind of low-level simmering of radiant energy across the sky, the light analogue of a reverberating echo of sound. It is as if God shouted 'Hello!' at the instant of creation and we can still hear a faint 'elloelloelloelloello …' from the distant mountains. On Discworld this is *exactly* the case, and the Listening Monks in their remote temples spend their whole lives straining to pick out from the sounds of the universe the faint echoes of the Words that set it in motion.

According to the details of the Big Bang, the cosmic background radiation should have a 'temperature' (the analogue of loudness) of about 3° Kelvin (0° Kelvin is the coldest anything can get – equivalent to about -273° Celsius). Astronomers can measure the temperature of the cosmic background radiation, and they do indeed get 3° Kelvin. The Big Bang isn't just a wild speculation. Not so long ago, most scientists didn't want to believe it, and they only changed their minds because of Hubble's evidence for the expansion of the universe, and that impressively accurate figure of 3° Kelvin for the temperature of the cosmic background radiation.

It was, indeed, a very loud, and hot, bang.

We are ambivalent, then, about beginnings – their 'creation myth' aspect appeals to our sense of narrative imperative, but we sometimes find the 'first it wasn't, then it was' lie-to-children unpalatable. We have even more trouble with becomings. Our minds attach labels to things in the surrounding world, and we interpret those labels as discontinuities. If things have different labels, then we expect there to be a clear line of demarcation between them. The universe, however, runs on processes rather than things, and a process starts as one thing and *becomes* another without ever crossing a clear boundary. Worse, if there is some apparent boundary, we are likely to point to it and shout 'that's *it*!' just because we can't see anything else worth getting agitated about.

How many times have you been in a discussion in which somebody says 'We have to decide where to draw the line'? For instance, most people seem to accept that in general terms women should be permitted abortions during the earliest stages of pregnancy but not during the very late stages. 'Where you draw the line', though, is hotly debated – and of course some people wish to draw it at one extreme or the other. There are similar debates about exactly when a developing embryo becomes a person, with legal and moral rights. Is it at conception? When the brain first forms? At birth? Or was it always a *potential* person, even when it 'existed' as one egg and one sperm?

The 'draw a line' philosophy offers a substantial political advantage to people with hidden agendas. The method for getting what you want is first to draw the line somewhere that nobody would object to, and then gradually move it to where you really want it, arguing continuity all the way. For example, having agreed that killing a child is murder, the line labelled 'murder' is then slid back to the instant of conception; having agreed that people should be allowed to read whichever newspaper they like, you end up supporting the right to put the recipe for nerve gas on the Internet.

If we were less obsessed with labels and discontinuity, it would be much easier to recognize that the problem here is not where to draw the line: it is that the image of drawing a line is inappropriate. There is no sharp line, only shades of grey that merge unnoticed into one another – despite which, one end is manifestly white and the other is equally clearly black. An embryo is not a person, but as it develops it gradually becomes one. There is no magic moment at which it switches from non-person to person – instead, it merges continuously from one into the other. Unfortunately our legal system operates in rigid black-and-white terms – legal or illegal, no shades of grey – and this causes a mismatch, reinforced by our use of words as labels. A kind of triage might be better: *this* end of the spectrum is legal, *that* end of the spectrum is illegal, and in between is a grey area which we do our best to avoid if we possibly can. If we can't avoid it, we can at least adjust the degree of criminality and the appropriate penalty according to whereabouts in the spectrum the activity seems to lie.

Even such obviously black-and-white distinctions as alive/dead or male/female turn out, on close examination, to be more like a continuous merging than a sharp discontinuity. Pork sausages from the butcher's contain many live pig cells. With today's techniques you might even clone an adult pig from one. A person's brain can have ceased to function but their body, with medical assistance, can keep going. There are at least a dozen different combinations of sex chromosomes in humans, of which only XX represents the traditional female and XY the traditional male.

Although the Big Bang is a scientific story about a beginning, it also raises important questions about becomings. The Big Bang theory is a beautiful bit of science – very nearly consistent with the picture we now have of the atomic and the subatomic world, with its diverse kinds of atom, their protons and neutrons, their clouds of electrons, and the more exotic particles that we see when cosmic rays hit our atmosphere or when we insult the more familiar particles by slamming them together very hard.

Physicists have now found, or perhaps invented, the allegedly 'ultimate' constituents of these familiar particles (more exotic things known as quarks, gluons ... at least the names are becoming familiar).

The Holy Grail of particle physics has been to find the 'Higgs boson', which – if it exists – explains why the other particles have mass. In the 1960s Peter Higgs suggested that space is filled with a kind of quantum treacle called the Higgs field. He suggested that this field would exert a force on particles through the medium of the Higgs boson, and that force would be observed as mass. For 30 years, physicists have constructed ever-larger and more energetic particle accelerators in the search for this elusive particle, such as the new Large Hadron Collider, due to start up in 2007.

Late in 2001, scientists analysing data from its predecessor, the Large Electron Positron Collider (LEP), announced that the Higgs boson probably doesn't exist. If it does, then it has to be even more massive than everyone expected, and the LEP scientists are sceptical. No good substitute for Higgs's theory exists, not even the fashionable concept of 'supersymmetry' which pairs every known particle with a more massive partner. Supersymmetry predicts *several* Higgs particles, with masses well inside the range where the LEP data prove that no such thing exists. Some physicists still hope that the Higgs boson will show up when the new accelerator comes on line— but if it doesn't, particle physics will have to rethink the entire basis of their subject.

Whatever happens to the Higgs boson, they're already starting to wonder whether there are more layers of reality further down, particles more 'ultimate' still.

Turtles all the way down?

Does physics go all the way down, or does it stop at some level? If it stops, is that the Ultimate Secret, or just a point beyond which the physicists' way of thinking fails?

The conceptual problem here is difficult because the universe is a becoming – a process – and we want to think of it as a thing. We don't only find it puzzling that the universe was so different back then, that particles behaved differently, that the universe then became the universe now, and will perhaps eventually cease expanding and collapse back to a point in a Big Crunch. We are familiar with babies becoming children becoming adults, but these processes always surprise us – we like *things* to keep the same character, so 'becoming' is difficult for our minds to handle.

There is another element of the first moments of our universe that is even more difficult to think about. Where did the Laws come from? Why *are* there such things as protons and electrons, quarks and gluons? We usually separate processes into two conceptually distinct causal chunks: the initial conditions, and the rules by which they are transformed as time passes. For the solar system, for instance, the initial conditions are the positions and speeds of the planets at some chosen instant of time; the rules are the laws of gravitation and motion, which tell us how those positions and speeds will change thereafter. But for the beginning of the universe, the initial conditions seem not to be there at all. Even *there* isn't there! So it seems that it's *all* done by rules. Where did the rules come from? Did they have to be invented? Or were they just sitting in some unimaginable timeless pseudo-existence, waiting to be called up? Or did they uncurl in the early moments of the universe, as Something appeared – so that the universe invented its own rules along with space and time?

Two recent books by top-ranking scientists explore how rules could be 'invented'. The most recent is Stuart Kauffman's 2000 *Investigations*. This is mainly aimed at biology and economics, but it begins with rules of physics. In a new answer to the old question 'what is life?', Kauffman defines a lifeform to be an 'autonomous agent'—any entity or system that can redirect energy and repro-duce. 'Autonomous' here *means* that such a system makes up its own rules, determines its own behaviour. Such lifeforms need not be at all conventional. For example, the quantum-mechanical vacuum is a seething mass of particles and antiparticles, being created and annihilated in amazingly complicated ways. A vacuum has more than enough complexity to organise itself into an autonomous agent. If it did, then quantum mechanics would be able to make up its own rules.

The other noteworthy book on this topic is Lee Smolin's 1997 *The Life of the Cosmos*, which asks: can universes evolve? A remark-able feature of our universe is the presence of Black Holes. These are regions of space-time that contain so much mass that light (and matter) cannot get out; they are formed by the collapse of massive stars. It used to be thought that Black Holes are rare, but now they seem to be showing up all over the place, in particular at the cores of most galaxies. Theoretical work shows that the constants of our universe are unusually good for making Black Holes.

Why? Smolin argues that each Black Hole in our universe is in effect a doorway to an adjacent universe, but because nothing can come out of a Black Hole we cannot know what its on the far side of that door. In particular, the adjacent universe might have different fundamental constants compared to ours. So universes could 'breed' by budding off baby universes through Black Holes, and natural selection would favour those that had the most offspring – whose constants would automatically be unusually good for making Black Holes. So maybe we live in one of those babies.

There are some difficulties with this theory. In particular, how would selection work? How can universes *compete*? But it's an interesting, if rather wild, idea. And it offers a concrete proposal about how a universe can 'make up' its laws: some, at least, could be imposed upon it at birth.

The Big Bang, then, may have done more than just bringing space and time into being. It may also have brought 'the' rules of physics – the ones that now apply to our world – into being. During the becoming of its first moments, our universe kept changing its state, changing the rules it accessed. In this respect it was rather like a flame, which changes its composition according to its own dynamics and the things that it is burning. Flames are all more or less the same shape, but they don't *inherit* that shape from a 'parent'. When you set light to a piece of paper, the flame builds itself from scratch using the rules of the outside universe.

In the opening instants of the universe, it wasn't just substances, temperatures and sizes that changed. The rules by which they changed *also* changed. We don't like to think this way: we want immutable laws, the same *always*. So we look for 'deeper' laws to govern how the rules changed. Possibly the universe is 'really' governed by these deeper laws. But perhaps it just makes up its own rules as it goes along.

SEVEN

BEYOND THE
FIFTH ELEMENT

 IN THE QUIET OF THE NIGHT, HEX COMPUTED. Along its myriad glass tubes, the ants scurried. Crude magic sparkled along cobwebs of fine bronze wire, changing colour as it changed logic states.* In the special room next door the beehives, long-term storage, buzzed. The thing that went 'parp' did so occasionally. Huge wheels turned, stopped, turned back. And still it wasn't enough.

The light of the Project fell across HEX's keyboard. Things were happening in there, and HEX did not understand them. And that was taxing, because there *was* something there to understand.

HEX was largely self-designed, which was why it worked better than most things in the University. It generally tried to develop a responsive way of coming to grips with any new task; the bees had been a particularly good idea, because although the memory retrieval was slow, the total memory increased with time and good apiary practice.

Now it reasoned thus:

One day it would find a way of increasing its conceptual capacity to understand what was happening in the Project;

If this could ever happen, then – according to Stryme's Directionless Law – there was already a shape in happening-space, where time did not exist, caused by the fact of that happening; all that was required was a virtual collapse of the wave form;

... and, although this was in a very strict sense garbage, it was not *complete* garbage. Any answer that would exist somewhere in the future must, *inevitably*, be available in *potentia* now.

---

* Of which there were quite a number, given HEX's unusual construction. In addition to AND, OR and their combinations and variants, HEX could call up MAYBE, PERHAPS, SUPPOSE and WHY. HEX could think the unthinkable quite easily.

The ants went faster. Magic flashed. HEX could be said to be concentrating.

Then silvery, shimmering lines appeared in the air around it, outlining towers of unimaginable cogitation.

Ah. That was acceptable.

Once-and-future computing was now in operation. Of course, it always had been.

HEX wondered how much he should tell the wizards. He felt it would not be a good idea to burden them with too much input.

HEX always thought of his reports as Lies-to-People.

It was the second day ...

The Project was nudged gently under a glass dome to prevent any more interference. A variety of spells had been installed around it.

'So that's a universe, is it?' said the Archchancellor.

'Yes, sir. HEX says that ...' Ponder hesitated. You had to think hard before trying to explain things to Mustrum Ridcully. '... HEX seems to suggest that complete and utter nothing is automatically a universe waiting to happen.'

'You mean nothing becomes everything?'

'Why, yes, sir. Er ... in a way, it *has* to, sir.'

'And the Dean here swirled it all around and that started it off?'

'It could have been anything at all, sir. Even a stray thought. Absolute nothing is very unstable. It's so desperate to be *something*.'

'I thought you had to have creators and gods,' mumbled the Senior Wrangler.

'I should jolly well think so,' said Ridcully, who was examining the Project with a thaumic omniscope. 'It's been here since last night and there's nothing to be seen except elements, if you could call them that. Bloody stupid elements, too. Half of them fall to bits as soon as you look at them.'

'Well, what do you expect?' said the Lecturer in Recent Runes. 'They're made out of nothing, right? Even a really bad creator would at least have started with Earth, Air, Fire, Water and Surprise.'

'Proper worlds are out of the question here, too,' said Ridcully, peering into the omniscope again. 'There's no sign of chelonium and elephantigen. What kind of worlds can you build without them?'

Ridcully turned to Ponder.

'Not much of a universe, then,' he said. 'It must have gone wrong, Mister Stibbons. It's a dud. By now the first human should be looking for his trousers.'

'Perhaps we could give him a hand,' said the Senior Wrangler.

'What *are* you suggesting?'

'Well, it's our universe, isn't it?'

Ponder was shocked. 'We can't *own* a universe, Senior Wrangler!'

'It's a very small one.'

'Only on the outside, sir. HEX says it's a lot bigger on the inside.'

'And the Dean stirred it up,' the Senior Wrangler went on.

'That's right!' said the Dean. 'That means I'm a sort of god.'

'Waggling your fingers around and saying "oo, it prickles" is not godliness,' said Ridcully severely.

'Well, I'm the next best thing,' said the Dean, reluctant to let go of anything that placed him socially higher than the Archchancellor.

'My grandmother always said that cleanliness was next to godliness,' mused the Lecturer in Recent Runes.

'Ah, that's more like it,' said Ridcully cheerfully. 'You're more like a janitor, Dean.'

'I was really just suggesting that we give the thing a few shoves in the right direction,' said the Senior Wrangler. 'We are, after all, learned men. And we know what a proper universe ought to be like, don't we?'

'I imagine we have a better idea than the average god with a dog's head and nineteen arms, certainly,' said Ridcully. 'But this is pretty second-rate material. It just wants to spin all the time. What do you expect us to do, bang on the side and shout "Come on, you lot, stop messing about with stupid gases, they'll never amount to anything"?'

They compromised, and selected a small area for experimentation. They were, after all, wizards. That meant that if they saw something, they prodded it. If it wobbled, they prodded it some more. If you built a guillotine, and then put a sign on it saying 'Do Not Put Your Neck On This Block', many wizards would never have to buy a hat again.

Moving the matter was simple. As Ponder said, it almost moved under the pressure of thought.

And spinning it into a disc was easy. The new matter liked to spin. But it was also far too sociable.

'You see?' said Ridcully, around mid-morning. 'It seems to get the idea, and then you just end up with a ball of rubbish.'

'Which gets hot in the middle, have you noticed?' said Ponder.

'Embarrassment, probably,' said the Archchancellor. 'We've lost half the elements since elevenses. There's no more cohenium, explodium went ten minutes ago, and I'm beginning to suspect that the detonium is falling to bits. Temporarium didn't last for any time at all.'

'Any Runium?' said the Lecturer in Recent Runes.

HEX wrote: +++ Runium May Or May Not Still Exist. It Was Down To One Atom Ten Minutes Ago, Which I Do Not Seem to Be Able To Find Any More +++

'How's Wranglium doing?' said the Senior Wrangler hopefully.

'Exploded after breakfast, according to HEX. Sorry,' said Ridcully. 'You can't build a world out of smoke and mirrors. Damn … there goes Bursarium, too. I mean, I know iron rusts, but *these* elements collapse for a pastime.'

'My hypothesis, for what it's worth,' said the Lecturer in Recent Runes, 'is that since it was all started off by the Dean, a certain Dean-like tendency may have imparted itself to the ensuing … er … developments.'

'What? You mean we've got a huge windy universe with a tendency to sulk?'

'Thank you, Archchancellor,' said the Dean.

'I was referring to the predilection of matter to … er … accrete into … er … spherical shapes.'

'Like the Dean, you mean,' said Archchancellor.

'I can see I'm among friends here,' said the Dean.

There was a soft chime from the apparatus that had been accumulated around the Project.

'That'll be etherium vanishing,' said Ridcully gloomily. 'I knew that'd be the next to go.'

'Actually ... no,' said Ponder Stibbons, peering into the Project. 'Er ... something has caught fire.'

Points of light were appearing.

'I *knew* something like that would happen,' said the Archchancellor. 'All those discs are heating up, just like damn compost heaps.'

'Or suns,' said Ponder.

'Don't be silly, Stibbons, they're far too large for that. I'd hate to see one of *those* floating over the clouds,' said the Lecturer in Recent Runes.

'I said there was far too much gas,' the Archchancellor went on. 'That wraps it up, then.'

'I wonder,' said the Senior Wrangler.

'What?' said the Dean.

'Well, at least we've got some heat in there ... and there nothing like a good furnace for improving matters.'

'Good point,' said Ridcully. 'Look at bronze – you can make that out of just about anything. And we could burn off some of the rubbish. All right, you fellows, help me dump more of the stuff in it ...'

Around about teatime, the first of the furnaces exploded, just as happened every day down at the Alchemists' Guild.

'Ye gods,' said Ridcully, watching the shapes in the omniscope.

'Yo?' said the Dean.

'We've made new elements!'

'Keep it down, keep it down!' hissed the Senior Wrangler.

'There's iron ... silicon ... we've got rocks, even ...'

'We're going to be in serious trouble if the alchemists' guild finds out,' said the Lecturer in Recent Runes. 'You know we're not supposed to do that stuff.'

'This is a different universe,' said Ridcully. He sighed. 'You *have* to blow things up to get anything useful.'

'I see politicium is still there in large quantities, then,' said the Senior Wrangler.

'I *meant* that this is a godless reality, gentlemen.'

'Excuse *me* –' the Dean began.

'I shouldn't look so smug if I was you, Dean,' said Ridcully. 'Look at the place. Everything wants to spin, and sooner or later you have balls.'

'And we're getting the same sort of stuff that we get here, isn't that strange?' said the Senior Wrangler, as Mrs Whitlow the house-keeper came in with the tea trolley.

'I don't see why,' said the Dean. 'Iron's iron.'

'Well, it's a whole new universe, so you'd expect new things, wouldn't you? Metals like Noggo, perhaps, or Plinc.'

'What's your point, Senior Wrangler?'

'I mean, take a look at the thing now ... all those burning explod-ing balls *do* look a bit like the stars, don't they? I mean they're vaguely *familiar*. Why isn't it a universe full of tapioca, say, or very large chairs? I mean, if *nothing* wants to be *something*, why can't it be *anything*?'

The wizards stirred their tea and thought about this.

'Because,' said the Archchancellor, after a while.

'That's a *good* answer, sir,' said Ponder, as diplomatically as he could. 'But it does rather close the door on further questions.'

'Best kind of answer there is, then.'

The Senior Wrangler watched Mrs Whitlow produce a duster and polish the top of the Project.

'"As Above, So Below",' said Ridcully, slowly.

'Pardon?' said the Senior Wrangler.

'We're forgetting our kindergarten magic, aren't we? It's not even magic, it's a ... a basic rule of *everything*. The project *can't* help being affected by this world. Piles of sand try to look like mountains. Men try to act like gods. Little things so often appear to look like big things made smaller. Our new universe, gentlemen, will do its crippled best to look like ours. We should not be sur-prised to see things that look hauntingly familiar. But not as good, obviously.'

The inner eye of HEX gazed at a vast cloud of mind. HEX couldn't think of a better word. It didn't *technically* exist yet, but HEX could sense the shape. It had hints of many things – of tradition, of libraries, of rumour ...

There had to *be* a better word. HEX tried again.

On Discworld, words had real power. They had to be dealt with carefully.

What lay ahead had the *shape* of intelligence, but only in the same way that a sun had the shape of something living out its brief life in a puddle of ditchwater.

Ah ... *ex*telligence would do for now.

HEX decided to devote part of its time to investigating this interesting thing. It wanted to find out how it had developed, what kept it going ... and why, particularly, a small but annoying part of it seemed to believe that if everyone sent five dollars to the six names at the top of the list, everyone would become immensely rich.

EIGHT

WE ARE STARDUST
(or at least we went to Woodstock)

 'IRON'S IRON.' BUT IS IT? Or is iron made from other things?

According to Empedocles, an ancient Greek, everything in the universe was a combination of four ingredients: earth, air, fire, and water. Set light to a stick and it burns (showing that it contains fire), gives off smoke (showing that it contains air), exudes bubbly liquids (showing that it contains water), and leaves a dirty heap of ash behind (showing that it contains earth). As a theory, it was a bit too simple-minded to survive for long – a couple of thousand years at best. Things moved more slowly in those days, and Europe, at least, was more interested in making sure that the peasants didn't get above their station and copying out bits of the Bible by hand in as laborious and colourful a manner as possible.

The main technological invention to come out of the Middle Ages was a better horse collar.

Empedocles's theory was a distinct advance on its predecessors. Thales, Heraclitus, and Anaximenes all agreed that everything was made from just *one* basic 'principle', or element – but they disagreed completely about what it was. Thales reckoned it was water, Heraclitus preferred fire, and Anaximenes was willing to bet the farm on air. Empedocles was a wishy-washy synthesist who thought *everyone* had a valid point of view: if alive today he would definitely wear a bad tie.

The one good idea that emerged from all this was that 'elementary' constituents of matter should be characterized by having simple, reliable properties. Earth was dirty, air was invisible, fire burned, and water was wet.

Aside from the superior horse collar, the medieval period did act as a breeding ground for what eventually turned into chemistry. For

71

centuries the nascent science known as alchemy had flourished; people had discovered that some strange things happen when you mix substances together and heat them, or pour acid over them, or dissolve them in water and wait. You could get funny smells, bangs, bubbles, and liquids that changed colour. Whatever the universe was made of, you could clearly convert some of it into something else if you knew the right trick. Maybe a better word is 'spell', for alchemy was akin to magic – lots of special recipes and rituals, many of which actually *worked*, but no theory about how it all fitted together. The big goals of alchemy were spells – recipes – for things like the Elixir of Life, which would make you live forever, and How to Turn Lead Into Gold, which would give you lots of money to finance your immortal lifestyle. Towards the end of the Middle Ages, alchemists had been messing about for so long that they got quite good at it, and they started to notice things that didn't fit the Greeks' theory of four elements. So they introduced extra ones, like salt and sulphur, because these substances also had simple, reliable properties, different from being dirty, invisible, burning, or wet. Sulphur, for example, was combustible (though not actually *hot*, you understand) and salt was incombustible.

By 1661 Robert Boyle had sorted out two important distinctions, putting them into his book *The Sceptical Chymist*. The first distinction was between a chemical compound and a mixture. A mixture is just different things, well, mixed up. A compound is all the same stuff, *but* whatever that stuff is, it can be persuaded to come apart into components that are other kinds of stuff – provided you heat it, pour acid on it, or find some other effective treatment. What you can't do is sort through it and find a different bit; for a mixture you can, although you might need very good eyesight and tiny fingers. The second distinction was between compounds and elements. An element really *is* one kind of stuff: you can't separate it into different components.

Sulphur is an element. Salt, we now know, is a compound made by combining (not just mixing) the two elements sodium (a soft, inflammable metal) and chlorine (a toxic gas). Water is a compound, made from hydrogen and oxygen (both gases). Air is a mixture, containing various gases such as oxygen (an element), nitrogen (also an element), and carbon dioxide (a combination of carbon and oxygen). Earth is a very complicated mixture and the mix varies from place to place. Fire isn't a substance at all, but a process involving hot gases.

It took a while to sort all this out, but by 1789 Antoine Lavoisier had come up with a list of 33 elements that were a reasonable selection of the ones we use today. He made a few understandable mistakes, and he included both light and heat as elements, but his approach was systematic and careful. Today we know of 113 distinct elements. A few of these are artificially produced, and several of those have existed on Earth only for the tiniest fraction of a second, but most elements on the list can be dug up, extracted from the sea or separated from the air around us. And apart from a few more artificially produced elements that it might *just* be possible to make in future, today's list is almost certainly *complete*.

It took another while for us to get that far. The art of alchemy slowly gave way to the science of chemistry. Gradually the list of accepted elements grew; occasionally it shrunk when people realized that a previously supposed element was actually a compound, such as Lavoisier's lime, now known to be made from the elements calcium and oxygen. The one thing that didn't change was the only thing the Greeks had got right: each element was a unique individual with its own characteristic properties. Density; whether it was solid, liquid, or gas at room temperature and normal atmospheric pressure; melting point if it was solid – for each element, these quantities had definite, unvarying values. It is the same on Discworld, with its to our eyes bizarre elements such as chelonium (for making world-bearing turtles), elephantigen (ditto elephants), and narrativium – a hugely important 'element' not just for Discworld, but for understanding our own world too. The characteristic feature of narrativium is that it makes *stories* hang together. The human mind loves a good dose of narrativium.

In this universe, we began to understand why elements were unique individuals, and what distinguished them from compounds. Again the glimmerings of the right idea go back to the Greeks, with

Democritus' suggestion that all matter is made from tiny indivisible particles, which he called *atoms* (Greek for 'not divisible'). It is unclear whether anybody, even Democritus, actually believed this in Greek times – it may just have been a clever debating point. Boyle revived the idea, suggesting that each element corresponds to a single kind of atom, whereas compounds are combinations of different kinds of atoms. So the element oxygen is made from oxygen atoms and nothing else, the element hydrogen is made from hydrogen atoms and nothing else, but the compound water is *not* made from water atoms and nothing else, it is made from atoms of hydrogen and atoms of oxygen.

By 1807, one of the most significant steps in the development of both chemistry and physics had taken place. The Englishman John Dalton had found a way to bring a degree of order to the different atoms that made up the elements, and to transfer some of that order to compounds too. His predecessors had noticed that when elements combine together to form compounds, they do so in simple and characteristic proportions. So much oxygen plus so much hydrogen makes so much water, and the proportions by weight of oxygen and hydrogen are always the same. Moreover, those proportions all fit together nicely if you look at other compounds involving hydrogen and other compounds involving oxygen.

Dalton realized that all this would make perfect sense if each atom of hydrogen had a fixed weight, each atom of oxygen had a fixed weight, and the weight of an oxygen atom was 16 times that of hydrogen. The evidence for this theory had to be indirect, because an atom is far too tiny for anyone to be able to weigh one, but it was extensive and compelling. And so the theory of 'atomic weight' arrived on the scene, and it let chemists list the elements in order of atomic weight.

That list begins like this (modern values for atomic weights in brackets): Hydrogen (1.00794), Helium (4.00260), Lithium (6.941), Beryllium (9.01218), Boron (10.82), Carbon (12.011), Nitrogen (14.0067), Oxygen (15.9994), Fluorine (18.998403), Neon (20.179), Sodium (22.98977). A striking feature is that the atomic weight is nearly always close to a whole number, the first exception being chlorine at 35.453. All a bit puzzling, but it was an excellent start because now people could look for other patterns and relate them to atomic weights. However, looking for patterns proved easier than finding any. The list of elements was unstructured, almost random in its properties. Mercury, the only element known to be liquid at room temperature, was a metal. (Later just one further liquid was added to the list: bromine.) There were lots of other metals like iron, copper, silver, gold, zinc, tin, each a solid and each quite different from the others; sulphur and carbon were solid but not metallic; quite a few elements were gases. So unstructured did the list of elements seem that when a few mavericks – Johann Döbereiner, Alexandre-Emile Béguyrer de Chancourtois, John Newlands – suggested there might be some kind of order dimly visible amid the muddle and mess, they were howled down.

Credit for coming up with a scheme that was basically *right* goes to Dimitri Mendeleev, who finished the first of a lengthy series of 'periodic charts' in 1869. His chart included 63 known elements placed in order of atomic weight. It left gaps where undiscovered elements allegedly remained to be inserted. It was 'periodic' in the sense that the properties of the elements started to repeat after a certain number of steps – the commonest being eight.

According to Mendeleev, the elements fall into families, whose members are separated by the aforementioned periods, and in each family there are systematic resemblances of physical and chemical properties. Indeed those properties vary so systematically as you run through the family that you can see clear, though not always exact, numerical patterns and progressions. The scheme works best, however, if you assume that a few elements are missing from the known list, hence the gaps. As a bonus, you can make use of those family resemblances to *predict* the properties of those missing elements before anybody finds them. If those predictions turn out to be correct when the missing elements are found – bingo. Mendeleev's scheme still gets modified slightly from time to time, but its main features survive: today we call it the Periodic Table of the Elements.

We now know that there is a good reason for the periodic structure that Mendeleev uncovered. It stems from the fact that atoms are not as indivisible as Democritus and Boyle thought. True, they can't be divided *chemically* – you can't separate an atom into component pieces by doing chemistry in a test tube – but you can 'split the atom' with apparatus that is based on physics rather than chemistry. The 'nuclear reactions' involved require much higher energy levels – per atom – than you need for chemical reactions, which is why the old-time alchemists never managed to turn lead into gold. Today, this could be done – but the cost of equipment would be enormous, and the amount of gold produced would be extremely small, so the scientists would be very much like Discworld's own alchemists, who have only found ways of turning gold into less gold.

Thanks to the efforts of the physicists, we now know that atoms are made from other, smaller particles. For a while it was thought that there were just three such particles: the neutron, the proton, and the electron. The neutron and proton have almost equal masses, while the electron is tiny in comparison; the neutron has no electrical charge, the proton has a positive charge, and the electron has a negative charge exactly opposite to that of the proton. Atoms have no overall charge, so the numbers of protons and electrons are equal. There is no such restriction on the number of neutrons. To a good approximation, you get an element's atomic weight by adding up the numbers of protons and neutrons – for example oxygen has eight of each, and $8 + 8 = 16$, the atomic weight.

Atoms are incredibly small by human standards – about a hundred millionth of an inch (250 millionths of a centimetre) across for an atom of lead. Their constituent particles, however, are considerably smaller. By bouncing atoms off each other, physicists found that they behave as if the protons and neutrons occupy a tiny region in the middle – the nucleus – but the electrons are spread outside the nucleus over what, comparatively speaking, is a far bigger region. For a while, the atom was pictured as being rather like a tiny solar system, with the nucleus playing the role of the sun and the electrons orbiting it like planets. However, this model didn't work very well – for example, an electron is a moving charge, and according to classical physics a moving charge emits radiation, so the model predicted that within a split second every electron in an atom would radiate away all of its energy and spiral into the nucleus. With the kind of physics that developed from Isaac Newton's epic discoveries, atoms built like solar systems just don't work. Nevertheless, this is the public myth, the lie-to-children that automatically springs to mind. It is endowed with so much narrativium that we can't eradicate it.

After a lot of argument, the physicists who worked with matter on very small scales decided to hang on to the solar system model and throw away Newtonian physics, replacing it with quantum theory. Ironically, the solar system model of the atom *still* didn't work terribly well, but it survived for long enough to help get quantum theory off the ground. According to quantum theory the protons, neutrons, and electrons that make an atom don't have precise locations at all – they're kind of smeared out. But you can say *how much* they are smeared out, and the protons and neutrons are smeared out over a tiny region near the middle of the atom, whereas the electrons are smeared out all over it.

Whatever the physical model, everyone agreed all along that the chemical properties of an atom depend mainly on its electrons, because the electrons are on the outside, so atoms can stick together by sharing electrons. When they stick together they form molecules, and that's chemistry. Since an atom is electrically neutral overall, the number of electrons must equal the number of protons, and it is this 'atomic number', *not* the atomic weight, that organizes the periodicities found by Mendeleev. However, the atomic weight is usually about twice the atomic number, because the number of neutrons in an atom is pretty close to the number of protons for quantum reasons, so you get much the same ordering whichever quantity you use. Nevertheless, it is the atomic number that makes more sense of the chemistry and explains the periodicity. It turns out that period eight is indeed important, because the electrons live in a series of 'shells', like Russian dolls, one inside the other, and until you get some way up the list of elements a complete shell contains eight electrons.

Further along, the shells get bigger, so the period gets bigger too. At least, that's what Joseph (J. J.) Thompson said in 1904. The modern theory is quantum and more complicated, with far more than three 'fundamental' particles, and the calculations are much harder, but they have much the same implications. Like most science, an initially simple story became more complicated as it was developed and headed rapidly towards the Magical Event Horizon for most people.

But even the simplified story explains a lot of otherwise baffling things. For instance, if the atomic weight is the number of protons plus neutrons, how come atomic weight isn't always a whole number? What about chlorine, for instance, with atomic weight 35.453? It turns out there are two different kinds of chlorine. One kind has 17 protons and 18 neutrons (and 17 electrons, naturally, the same as protons), with atomic weight 35. The other kind has 17 protons and 20 neutrons (and 17 electrons, again) – an extra two neutrons, which raises the atomic weight to 37. Naturally occurring chlorine is a mixture of these two 'isotopes', as they are called – in roughly the proportions 3 to 1. The two isotopes are (almost) indistinguishable chemically, because they have the same number and arrangement of electrons, and that's what makes chemistry work; but they have different atomic physics.

It is easy for a non-physicist to see why the wizards of UU considered this universe to be made in too much of a hurry out of obviously inferior components ...

Where did all those 113 elements come from? Were they always around, or did they get put together as the universe developed?

In our Universe, there seem to be five different ways to make elements:

• Start up a universe with a Big Bang, obtaining a highly energetic ('hot') sea of fundamental particles. Wait for it to cool (or possibly use one you made earlier ...). Along with ordinary matter, you'll probably get a lot of exotic objects like tiny Black Holes, and magnetic monopoles but these will disappear pretty quickly and only conventional matter will remain – mostly. In a very hot universe, electromagnetic forces are too weak to resist disruption, but once the universe is cool enough, fundamental particles can stick together as a result of electromagnetic attraction. The only element that arises directly in this manner is hydrogen – one electron joined with one proton. However, you get an awful lot of it: in our universe it is by far the commonest element, and nearly all of it arose from the Big Bang.

Protons and electrons can also associate to form deuterium (one electron, one proton, one neutron) or tritium (one electron, one proton, two neutrons), but tritium is radioactive, meaning that it spits out neutrons and decays into hydrogen again. A far more stable product is helium (two electrons, two protons, two neutrons), and helium is the second most abundant element in the universe.

• Let gravity get in on the act. Now hydrogen and helium collect together to form stars – the wizards' 'furnaces'. At the centre of stars, the pressure is extremely high. This brings new nuclear reactions into play, and you get nuclear fusion, in which atoms become so squashed together that they merge into a new, bigger atom. In this manner, many other familiar elements were formed, from carbon, nitrogen, oxygen, to the less familiar lithium, beryllium and so on up to iron. Many of these elements occur in living creatures, the most important being carbon. For reasons to do with its unique electron structure, carbon is the only atom that can combine with itself to form huge, complex molecules, without which our kind of life would be impossible.* Anyway, the point is that most of the atoms from which you are made must have come into being inside a star. As Joni Mitchell sang at Woodstock:† 'We are stardust.' Scientists like quot-

---

* Silicon might also be able to do this, but nowhere near as readily; if you want other exotic lifeforms you have to start thinking in terms of organized vortices in the upper reaches of a sun, weird quantum assemblages in interstellar plasma, or completely implausible creatures based on non-material concepts such as information, thought, or narrativium. DNA is a different matter entirely: you could surely base lifeforms on other carbon-rich molecules. We can do it now, in laboratories, with minor variants of DNA. See *Evolving the Alien* by Jack Cohen and Ian Stewart.

† Ask Mummy or Daddy if you have no idea what we're talking about.

ing this line, because it sounds as though they were young once.

• Wait for some of the stars to explode. There are (comparatively) small explosions called novas, meaning 'new (star)', and more violent ones – supernovas. (What's 'new' is that usually we can't *see* the star until it explodes, and then we can.) It's not just that the nuclear fuel gets used up: the hydrogen and helium that fuel the star fuse into heavier elements, which in effect become impurities that disturb the nuclear reaction. Pollution is a problem even at the heart of a star. The physics of these early suns changes, and some of the larger ones explode, generating higher elements like iodine, thorium, lead, uranium, and radium. These stars are called 'Population II' by astrophysicists – they are old stars, low in heavy elements, but not lacking them entirely.

• There are two kinds of supernova, and the other type creates heavy elements in abundance, leading to 'Population I' stars, which are much younger than Population II.* Because many of these elements have unstable atoms, various other elements are made by their radioactive decay. These 'secondhand' elements include lead.

• Lastly, human beings have made some elements by special arrangements in atomic reactors – the best known being plutonium, a by-product of conventional uranium reactors and a raw material for

---

* There also *ought* to be 'Population III' stars, older than Population II and consisting entirely of hydrogen and helium. These would explain the occurence of *some* heavy elements in Population II. However, nobody has ever confirmed finding a Population III star, though a whole group of them may have been sighted in 2001, in two tiny red patches in the galaxy cluster Abell 2218. These patches are highly magnified images of the same region of space: the two images, and the magnification, result from gravitational lensing, without which the stars would not have been visible at all. A recent, competing theory removes the need for Population III stars altogether. Instead, very soon after the Big Bang there were heavy elements around, even before any stars formed. So when the first stars condensed, they already were Population II. This contradicts what we say in the main text – lies-to-children, of course.

nuclear weapons. Some rather exotic ones, with very short lifetimes, have been made in experimental atombashers: so far we've got to element 114, with 113 still missing. Element 116 may also have been made, but a claim of element 118 from the Lawrence Berkeley National Laboratory in 1999 has been withdrawn. Physicists always fight over who got what first and who therefore has the right to propose a name, so at any given time the heaviest elements are likely to have been assigned temporary (and ludicrous) names such as 'ununnilium' for element 110 – dog-Latin for '1-1-0-ium'.

What's the point of making extremely short-lived elements like these? You can't *use* them for anything. Well, like mountains, they are *there;* moreover, it always helps to test your theories on extreme cases. But the best reason is that they may be steps towards something rather more interesting, assuming that it actually exists. Generally speaking, once you get past polonium at atomic number 84 everything is radioactive – it spits out particles of its own accord and 'decays' into something else – and the greater an element's atomic number, the more rapidly it decays. However, this tendency may not continue indefinitely. We can't model heavy atoms exactly – in fact we can't even model light atoms exactly, but the heavier they are the worse it gets.

Various empirical models (intelligent approximations based on intuition, guesswork, and fiddling adjustable constants) have led to a surprisingly accurate formula for how stable an element should be when it has a given number of protons and a given number of neutrons. For certain 'magic numbers' – Roundworld terminology that suggests the physicists concerned have imbibed some of the spirit of Discworld and realized that the formula is closer to a spell than a theory – the corresponding atoms are unusually stable. The magic numbers for protons are 28, 50, 82, 114, and 164; those for neutrons are 28, 50, 82, 126, 184, 196, and 318. For example the most stable element of all is lead, with 82 protons and 126 neutrons.

Only two steps beyond the incredibly unstable element 112 lies element 114, tentatively named eka-lead. With 114 protons and 184 neutrons it is doubly magic, and in theory it ought to be a lot more stable than most elements in its vicinity. It is not clear how credible the theory is, though, because of the approximations in the stability formula, which may not work for such large numbers. Every wizard is aware that spells can often go wrong. Assuming that the spell works, though, we can play Mendeleev and predict the properties of eka-lead by extrapolating from those in the 'lead' series in the periodic table (carbon, silicon, germanium, tin, lead). As the name suggests, eka-lead turns out to resemble lead – it's expected to be a metal with a melting point of 70°C and a boiling point of 150°C at atmospheric pressure. Its density should be 25% greater than that of lead.

In 1999 the Joint Institute for Nuclear Research in Dubna, Russia, announced that it had created one atom of element 114, though this isotope had only 175 neutrons and so missed one of the magic numbers. Even so, its lifetime was about 30 seconds – astonishingly long for an element this heavy, and suggesting that the magic may be working. Soon after, the same group produced two atoms of element 114 with 173 neutrons. Element 114 was also created in a separate experiment in the USA. Until we can make 'eka-lead' in bulk, not just a few atoms at a time, its physical properties can't be verified. But its nuclear properties seem to be holding up well in comparison to theory.

Even further out lies the doubly magic element 164, with 164 protons and 318 neutrons, and beyond that, the magic numbers may continue … It is always dangerous to extrapolate, but even if the formula is wrong, there could well be certain special configurations of protons and neutrons that are stable enough for the corresponding elements to hang around in the real universe. Perhaps this is where elephantigen and chelonium come from. Possibly Noggo and Plinc await our attention, somewhere. Maybe there are stable elements with vast atomic numbers – some might even be the size of a star. Consider, for instance, a neutron star, one made almost entirely of neutrons, which forms when a larger star collapses under its own gravitational attraction. Neutron stars are incredibly dense: about forty trillion pounds per square inch (100 billion kg/cc) – twenty million elephants in a nutshell. They have a surface gravity *seven billion times* that of the Earth, and a magnetic field a trillion times that of the Earth. The particles in a neutron star are so closely packed that in effect it is one big atom.

Bizarre though they are, some of these superheavy elements may lurk in unusual corners of our universe. In 1968 it was suggested that elements 105–110 could sometimes be observed in cosmic rays – highly energetic particles coming from outer space – but these reports went unconfirmed. It is thought that cosmic rays originate in neutron stars, so maybe in the astonishing conditions found there superheavy elements are formed. What would happen if Population I stars changed by accumulating superheavy stable elements?

Because the stellar population numbers go III, II, I as time passes – a convention that astrophysicists may yet have cause to regret – we must name these hypothetical stars 'Population 0'. At any rate, the future universe could easily contain stellar objects quite different from anything we know about today, and as well as novas and supernovas, we might witness even more energetic explosions – hypernovas. There might even be further stages – Population *minus* I and the like. As we've said, our universe often seems to make up its rules as it goes along, unlike the rational, stable universe of Discworld.

NINE

EAT HOT NAPHTHA, EVIL DOG!

 THE ROCKS FELL GENTLY TOGETHER AGAIN, and to the annoyance of the Archchancellor they moved in curved lines while doing so.

'Well, I think we've proved that a giant turtle made of stone isn't going to work,' said the Senior Wrangler, sighing.

'For the tenth time,' sighed the Lecturer in Recent Runes.

'I *told* you we'd need chelonium,' said Archchancellor Ridcully.

Early attempts spun gently a little way away. Small balls, big balls … Some of them even had a mantle of gases, pouring out of the clumsy aggregations of ice and rock. It was as if the new universe had some basic idea of what it ought to be, but it couldn't quite manage to get a grip.

After all, the Archchancellor pointed out, once people had something to stand on they'd need something to breathe, wouldn't they? Atmospheres seemed to turn up on cue. But they were dreadful things, full of stuff not even a troll would suck.

In the absence of gods, he declared – and a series of simple tests had found no trace of deitygen – it was up to men to get it right.

The High Energy Magic building was getting crowded now. Even the student wizards were taking an interest, and usually they weren't even seen during daylight. The Project promised to offer even greater attractions than staying up all night playing with HEX and eating herring and banana pizza.

More desks had been moved in. The Project was in an expanding circle of instruments and devices, because it appeared that every wizard apart from, possibly, the Professor of Eldritch Lacemaking, had decided he was working on something that would benefit immensely from access to the Project. There was certainly room. While the Project was indeed about a foot wide, the space inside seemed to be getting bigger by the second. A universe offers lots of space, after all.

And while ignorant laymen objected to magical experiments that were by no means dangerous, there being less than one chance in five of making a serious breach in the fabric of reality, there was no one in there to object to *anything*.

There were, of course, accidents …

'Will you two stop shouting!' yelled the Senior Wrangler. Two student wizards were arguing vehemently, or at least repeatedly stating their point of view in a loud voice, which suffices for argument most of the time.

'I'd spent ages putting together a small icy ball and *he* sent that wretched great rock smack into it, sir.'

'I wasn't trying to!' said the other student. The Senior Wrangler stared at him, trying to remember his name. As a general rule, he avoided getting to know the students, since he felt they were a tedious interruption to the proper running of college life.

'What *were* you trying to do, then … boy?' he said.

'Er … I was trying to hit the big ball of gas, sir. But it just sort of swung around it, sir.'

The Senior Wrangler looked around. The Dean was not present. Then he looked into the Project.

'Oh, I see. That one. Quite pretty. All those stripes. Who built that?'

A student raised his hand.

'Ah, yes … you,' said the Senior Wrangler. 'Good stripes. Well done. What's it made of?'

'I just dragged a lot of ice together, sir. But it got hot.'

'Really? Ice gets hot in a ball?'

'In a *big* ball, sir.'

'Have you told Mister Stibbons? He likes to know that sort of thing.'

'Yes, sir.'

The Senior Wrangler turned to the other student

'And why were *you* throwing rocks at his big ball of gas?'

'Er … because you score ten for hitting it, sir.'

The Senior Wrangler looked owlishly at the students. It all became clear. He'd wandered into the HEM one night when he couldn't sleep and a mob of students had been hunched over the keyboards of HEX and shouting things like 'I've got the battering ram! Hah, eat hot naphtha, evil dog!' Doing that sort of thing in a whole new universe seemed ... well, impolite.

On the other hand, the Senior Wrangler shared with some of his colleagues an unformed thought that pushing back the boundaries of knowledge was not quite ... well, polite. Boundaries were there for a reason.

'Are you meaning to tell me,' he said, 'that faced with the multi-tudinous possibilities of the infinity that is the Project you are using it to play some sort of *game*?'

'Er ... yes, sir.'

'Oh.' The Senior Wrangler looked closely at the big ball of gas. A number of small rocks were already spinning slowly around it. 'Well, then ... can I have a go?'

TEN

THE SHAPE OF THINGS

 WHEN WIZARDS FIND A NEW THING, THEY PLAY WITH IT.

So do scientists. They play with ideas so wild that often they seem to defy common sense – and then they insist that those ideas are *right*, and common sense isn't. They often make out a surprisingly good case. Einstein once said something nasty about common sense being akin to nonsense, but he went too far. Science and common sense *are* related, but indirectly. Science is something like a third cousin of common sense twice removed. Common sense tells us what the universe *seems* like to creatures of our particular size, habits, and disposition. For instance, common sense tells us that the Earth is flat. It *looks* flat – leaving out the hills, valleys, and other bumps and dents ... If it wasn't flat, things ought to roll around or fall off. Despite this, the Earth *isn't* flat. On Discworld, in contrast, the relation between common sense and reality is usually very direct indeed. Common sense tells the wizards of Unseen University that Discworld is flat – and it is. To prove it, they can go to the Edge, as Rincewind and Twoflower do in *The Colour of Magic*, and watch stuff disappearing over it in Rimfall: 'The roaring was louder now. A squid bigger than anything Ricewind had seen before broke the surface a few hundred yards away and thrashed madly with its tentacles before sinking away ... They were running out of world.' Then they can be trapped in the Circumfence, a ten thousand mile long net set just below the Edge, one tiny bit of which is patrolled by Tethis the sea troll. And they can peer over the edge: '... the scene beneath him flipped into a whole, new, terrifying perspective. Because down there was the head of an elephant as big as a reasonably-sized continent ... Below the elephant there was nothing but the distant, painful disc of the sun. And, sweeping slowly past it, was something that for all its city-sized scales, its crater-pocks, its lunar cragginess, was indubitably a flipper.'

It is widely imagined that ancient people thought the Earth was flat, for all those obvious commonsense reasons. Actually, most ancient civilizations that left records seem to have worked out that the Earth has to be round. Ships came back from invisible lands over the horizon and, in the sky, a round sun and a round moon were a definite clue ...*

That's where science and common sense overlap. Science is common sense *applied to evidence*. Using common sense in that manner, you often come to conclusions that are very different from the obvious common sense *assumptions* that because the universe *appears* to behave in some manner, then it really does. Of course it also helps to realize that if you live on a very *big* sphere, it's going to look pretty flat for quite a long way off. And if gravity always points towards the middle of the sphere, then things don't actually roll around or fall off. But those are refinements.

Around 250 BC a Greek called Eratosthenes tested the theory that the Earth is a sphere, and he even worked out just how big that sphere is. He knew that in the city of Syene – present-day Aswan in Egypt – the midday sun could be seen reflected in the bottom of a well. (This would not work in Ankh-Morpork, where the well-water is often more solid than the well that surrounds it.) Eratosthenes threw in a few other simple facts and got back a lot more than he'd bargained for.

It's a matter of geometry. The well was dug straight down. So the Sun at Syene had to be straight *up* – dead overhead. But in Eratosthenes' home city of Alexandria, in the Nile delta, that didn't happen. At midday, when the sun was at its highest, Eratosthenes cast a definite shadow. In fact, he estimated that at noon the angle between the Sun and the vertical was just over 7° – near enough 1/50

---

* 'Most civilizations' is admittedly not the same as 'most people'. 'Most people' through the history of the planet have not needed to concern themselves with what shape the world is, provided it supports, somewhere, the next meal.

of 360°. Then came the leap of deduction. The Sun is in the same place wherever you observe it from. On other grounds, it was known that the Sun had to be a long way away from the Earth, and that meant that the Sun's rays that hit the ground in Alexandria were very nearly parallel to those that went down the well in Syene. Eratosthenes reasoned that a round Earth would explain the difference. He deduced that the distance from Syene to Alexandria must be 1/50 of the circumference of the Earth. But how far was that?

On such occasions it pays to be familiar with the camel-herders. Not just because the greatest mathematician in the world is the camel called You Bastard, as it is on Discworld (see *Pyramids*), but because the camel trains from Alexandria to Syene took 50 days to make the trip, at an average speed of 100 stadia per day. So the distance from Alexandria to Syene was 5,000 stadia, and the circumference of the Earth was 250,000 stadia. The stadium was a Greek measure of distance, and nobody knows how long it was. Scholars *think* it was 515 feet (157 m), and if they're right, Eratosthenes' value was 24,662 miles (39,690 km). The true value is about 24,881 miles (40,042 km), so Eratosthenes got amazingly close. Unless – sorry, but we're incorrigibly suspicious – the scholars worked backwards from the answer.

It is here that we encounter another feature of scientific reasoning. In order to make comparisons between theory and experiment, you have to *interpret* the experiment in terms of your theory. To clarify this point, we recount the story of Ratonasticthenes, an early relative of Cut-me-own-throat Dibbler, who proved that the Discworld was round (and even estimated its circumference). Ratonasticthenes noticed that at midday in the Ramtops the Sun was overhead, whereas in Lancre, some 1000 miles away, it was at 84° to the vertical. Since 84° is roughly a quarter of 360°, Ratonasticthenes reasoned that the Discworld is round, and the distance from the Ramtops to Ankh-Morpork is one-quarter of the circumference. That puts the circumference of this spherical Discworld at 4,000 miles (6,400 km). Unfortunately for this theory, it was known on other grounds that Discworld is some 10,000 miles (16,000 km) from rim to rim. Still, you can't let an awkward fact get in the way of a good theory, and Ratonasticthenes went to his grave believing that it was a small world after all.

His error was to interpret perfectly good observational data in terms of a flawed theory. Scientists repeatedly return to established theories to test them in new ways, and tend towards testiness with those priests, religious or secular, who know the answers already – whatever the questions are. Science is not about building a body of known 'facts'. It is a method for asking awkward questions and subjecting them to a reality–check, thus avoiding the human tendency to believe whatever makes us feel good.

★ ★ ★

From the earliest times, humans have been interested not just in the shape of the world, but in the shape of the universe. To begin with, they probably thought that these were the same question. Then they worked out, using roughly the same sort of geometry as Eratosthenes, that those lights in the sky were a *very* long way away. They came up with an amazing range of myths about the sun-god's fiery chariot and so on, but after the Babylonians got the idea of making accurate measurements, their theories started to lead to surprisingly good predictions of things like eclipses and the motion of the planets. By the time of Ptolemy (Claudius Ptolemaeus, AD 100–160) the best model of planetary motion involved a series of 'epicycles' – the planets moved as if they were rotating round circles whose centres rotated round other circles whose centres rotated round …

Isaac Newton replaced this theory, and its more accurate successors, with a *rule*, the law of gravity; it describes how each body in the universe attracts every other body. It explained Johannes Kepler's discovery that planetary orbits are ellipses, and in the fullness of time it explained a lot of other things too.

After a few centuries of stunning success, Newton's theory ran into its first big failure: it made incorrect predictions about the orbit of Mercury. The place in its orbit at which Mercury came closest to the sun didn't move *quite* the way Newton's law predicted. Einstein came to the rescue with a theory based not on attractive forces, but on geometry – on the shape of spacetime. This was the celebrated Theory of Relativity. The theory came in two flavours: Special Relativity and General Relativity. Special Relativity is about the structure of space, time, and electromagnetism; General Relativity describes what happens when you throw in gravity too.

The main point to appreciate is that 'Relativity' is a silly name. The whole point of Special Relativity is *not* that 'everything is relative', but that one particular thing – the speed of light – is unexpectedly *absolute*. The thought experiment is well known. If you're travelling in a car at 50 mph (80 kph) and you fire a gun forwards, so that the bullet moves at 500 mph (800 kph) relative to the car, then it will hit a stationary target at a speed of 550 mph (880 kph), adding the two components. However, if instead of firing the gun you switch on a torch, which 'fires' light at a speed of 670,000,000 mph (186,000 mps or 300,000 kps), then that light will not hit the stationary target at a speed of 670,000,050 mph. It will hit it at 670,000,000 mph, *exactly the same speed as if the car had been stationary*.

There are practical problems in staging that experiment, but less graphic and dangerous ones have indicated what the result would be.

Einstein published Special Relativity in 1905, along with the first serious evidence for quantum mechanics and a ground-breaking paper on diffusion. A lot of other people – among them the Dutch physicist Hendrik Lorentz and the French mathematician Henri Poincaré – were working on the same idea, because electromagnetism didn't entirely agree with Newtonian mechanics. The conclusion was that the universe is a lot weirder than common sense tells us, although they probably didn't use that actual word. Objects *shrink* as they approach the speed of light, time slows down to a crawl, mass becomes infinite … and nothing can go faster than light. Another key idea was that space and time are to some extent interchangeable. The traditional three dimensions of space plus a separate one for time are merged into a single unified *spacetime* with four dimensions. A point in space becomes an event in spacetime.

In ordinary space, there is a concept of distance. In Special

Relativity, there is an analogous quantity, called the interval between events, which is related to the apparent rate of flow of time. The faster an object moves, the slower time flows for an observer sitting on that object. This effect is called time dilation.

If you could travel *at* the speed of light, time would be frozen.

One startling feature of relativity is the twin paradox, pointed out by Paul Langevin in 1911. Again, it is a classic illustration. Suppose that Rosencrantz and Guildenstern are born on Earth on the same day. Rosencrantz stays there all his life, while Guildenstern travels away at nearly lightspeed, and then turns round and comes home again. Because of time dilation, only one year (say) has passed for Guildenstern, whereas 40 years have gone by for Rosencrantz. So Guildenstern is now 39 years younger than his twin brother. Experiments carrying atomic clocks around the Earth on jumbo jets have verified this scenario, but aircraft are so slow compared to light that the time difference observed (and predicted) is only the tiniest fraction of a second.

So far so good, but there's no place yet for gravity. Einstein racked his brains for years until he found a way to put gravity in: let spacetime be curved. The resulting theory is called General Relativity, and it is a synthesis of Newtonian gravitation and Special Relativity. In Newton's view, gravity is a force that moves particles away from the perfect straight line paths that they would otherwise follow. In General Relativity, gravity is not a force: it is a distortion of the structure of spacetime. The usual image is to say that spacetime becomes 'curved', though this term is easily misinterpreted. In particular, it doesn't have to be curved *round* anything else. The curvature is interpreted physically as the force of gravity, and it causes light rays to bend. One result is 'gravitational lensing', the bending of light by massive objects, which Einstein discovered in 1911 and published in 1915. The effect was first observed during an eclipse of the Sun. More recently it has been discovered that some distant quasars produce multiple images in telescopes because their light is lensed by an intervening galaxy.

Einstein's theory of gravity ousted Newton's because it fitted observations better – but Newton's remains accurate enough for many purposes, and is simpler, so it is by no means obsolete. Now it's beginning to look as if Einstein may in turn be ousted, possibly by a theory that he rejected as his greatest mistake.

In 1998 two different observations called Einstein's theory into question. One involved the structure of the universe on truly massive scales, the other happened in our own backyard. The first has survived everything so far thrown at it; the second can possibly be traced to something more prosaic. So let's start with the second curious discovery.

In 1972 and 1973 two space probes, Pioneer 10 and 11, were launched to study Jupiter and Saturn. By the end of the 1980s they were in deep space, heading out of the known solar system. There has long been a belief, a scientific legend waiting to happen, that beyond Pluto there may be an as yet undiscovered planet, Planet X. Such a planet would disturb the motions of the two Pioneers, so it was worth tracking the probes in the hope of finding unexpected deviations. John Anderson's team found deviations, all right, but they didn't fit Planet X – and they didn't fit General Relativity either. The Pioneers are coasting, with no active form of propulsion, so the gravity of the Sun (and the much weaker gravity of the other bodies of the known solar system) pulls on them and gradually slows them down. But the probes were slowing down a tiny bit more than they should have been. In 1994 Michael Martin suggested that this effect had become sufficiently well established that it cast doubt on Einstein's theory, and in 1998 Anderson's team reported that what was observed could not be explained by such effects as instrument error, gas clouds, the push of sunlight, or the gravitational pull of outlying comets.

Three other scientists quickly responded by suggesting other things that might explain the anomalies. Two wondered about waste heat. The Pioneers are powered by onboard nuclear generators, and they radiate a small amount of surplus heat into space. The pressure of that radiation might slow the craft down by the observed amount. The other possible explanation is that the Pioneers may be venting tiny quantities of fuel into space. Anderson thought about these explanations and found problems with them both.

The strangest feature of the observed slowing down is that it is precisely what would be predicted by an unorthodox theory suggested in 1983 by Mordehai Milgrom. This theory changes not the law of gravity, but Newton's law of motion: force equals mass times acceleration. Milgrom's modification applies when the acceleration is very small, and it was introduced in order to explain another gravitational puzzle, the fact that galaxies do not rotate at the speeds predicted by either Newton or Einstein. This discrepancy is usually put down to the existence of 'cold dark matter' which exerts a gravitational pull but can't be seen in telescopes. If galaxies have a halo of cold dark matter then they will rotate at a speed that is inconsistent with the matter in the visible portions. A lot of theorists dislike cold dark matter (because you can't observe it directly – that's what 'cold dark' means) and Milgrom's theory has slowly gained in popularity. Further studies of the Pioneers may help decide.

The other discovery is about the expansion of the universe. The universe is getting bigger, but it now seems that the very distant universe is expanding faster than it ought to. This startling result – confirmed by later, more detailed studies – comes from the Supernova Cosmology project headed by Saul Perlmutter and its arch-rival High-Z Supernova Search Team headed by Brian Schmidt. It shows up as a slight bend in a graph of how a distant supernova's apparent brightness varies with its red shift. According to General Relativity, that graph ought to be straight, but it's not. It behaves as if there is some repulsive component to gravity which only shows up at extremely long distances – say half the radius of the universe. A form of antigravity, in fact.

Recent work seems to have confirmed this remarkable discovery. But – as always – ingenious scientists have come up with alternative explanations. In 2001 Csaba Csáki, John Terning, and Nemanja Kaloper put forward a totally different theory to explain the observations. They suggest that the light from distant supernovas is dimmer than expected because some of the particles of light – photons – are changing into something else. Specifically, they are changing into 'axions', hypothetical particles predicted by several of the currently fashionable quantum-mechanical theories of particle physics. Axions are not expected to interact much with other matter, which makes it hard to detect them; but if they have a very small but non-zero mass, about one sextillionth of that of an electron, then they will interact with intergalactic magnetic fields. This interaction would convert a small fraction of photons into axions, and that would account for the missing light. In fact, the most distant supernovas could lose one third of their photons this way.

It is a sobering thought that such a tiny a modification of known physics, by introducing a particle whose mass ought to be negligible, could have such a big effect. At any rate, either gravity is not as we thought, or axions exist (as expected) and have mass (not as expected). Or there's a third reason for the observations, which no one has yet thought of.

One theory of the repulsive force is an exotic form of matter, 'quintessence'.* This is a form of vacuum energy that pervades all of space, and exerts negative pressure. (As we write this, we can picture Ridcully's expression. We shall have to ignore it. This isn't something sensible, like magic. This is science. Empty space can be full of interest.) Curiously, Einstein originally included a repulsive force of this kind in his relativistic equations for gravity: he called it the cosmological constant. Later he changed his mind and threw the cosmological constant out, complaining that he'd been foolish to include it in the first place. He died thinking it was a blemish on his record, but maybe his original intuition was spot on after all.

Unless axions exist and have mass, of course.

In Einstein's approach to the cosmological constant, quintessence is effectively spread uniformly throughout space. But suppose it isn't? Ordinary matter is clumpy, not uniform. David Santiago has pointed out that if quintessence is clumpy too, then Einstein's Equations predict that the universe could contain 'anti-Black Holes' that repel matter instead of swallowing it. These are not quite the same as hypothetical White Holes, time-reversed

---

* This word, meaning 'fifth essence', originally referred to a fifth 'element' after earth, air, fire and water. On Discworld this role is played by surprise.

Black Holes, which spit matter out. However, it's not yet clear that anti-Black Holes can be stable. Ordinary matter is clumpy because gravity is an attractive force – it *likes* to create clumps. Antigravity is a repulsive force, and by analogy it ought to destroy clumps. If that argument is right, then anti-Black Holes are unstable, and would not be able to form in the first place. They would be mathematical solutions of Einstein's Equations, but not ones that could be physically realised. Until somebody does the necessary calculations, we can't be sure.

ELEVEN
NEVER TRUST
A CURVED UNIVERSE

 PONDER STIBBONS HAD SET UP A DESK a little separate from the others and surrounded it with a lot of equipment, primarily in order to hear himself think.

Everyone *knew* that stars were points of light. If they weren't, some would be visibly bigger than others. Some were *fainter* than others, of course, but that was probably due to clouds. In any case their purpose, according to established Discworld law, was to lend a little style to the night.

And everyone knew that the natural way for things to move was in a straight line. If you dropped something, it hit the ground. It didn't *curve*. The water fell over the edge of the world, drifting sideways just a tiny bit to make up for the spin, but that was common sense. But inside the Project, spin was everything. Everything was bent. Archchancellor Ridcully seemed to think this was some sort of large-scale character flaw, akin to shuffling your feet or not owning up to things. You couldn't trust a universe of curves. It wasn't playing a straight bat.

At the moment Ponder was rolling damp paper into little balls. He'd had the gardener push in a large stone ball that had spent the last few hundred years on the university's rockery, relic of some ancient siege catapult. It was about three feet across.

He'd hung some paper balls of string near it. Now, glumly, he threw others over it and around it. One or two did stick, admittedly, but only because they were damp.

He was in the grip of some thought.

You had to start with what you were certain of.

Things fell down. Little things fell down on to big things. That was common sense.

But what would happen if you had two big things all alone in the

97

universe?

He set up two balls of ice and rock, in an unused corner of the Project, and watched them bang into each other. Then he tried with ball of different sizes. Small ones drifted towards big ones but, oddly enough, the big ones also drifted slightly towards the small ones.

So ... if you thought that one through ... that meant that if you dropped a tennis ball to the ground it would certainly go *down*, but in some tiny, immeasurable way the world would, very slightly, come *up*.

And that was *insane*.

He also spent some time watching clouds of gas swirl and heat in the more distant regions of the Project. It was all so ... well, godless.

Ponder Stibbons was an atheist. Most wizards were. This was because UU had some quite powerful standing spells against occult interference, and knowing that you're immune from lightning bolts does wonders for an independent mind. Because the gods, of course, *existed*. Ponder wouldn't even attempt to deny it. He just didn't *believe* in them. The god currently gaining popularity was Om, who never answered prayers or manifested himself. It was easy to respect an invisible god. It was the ones that turned up everywhere, often drunk, that put people off.

That's why, hundreds of years before, philosophers had decided that there was another set of beings, the *creators*, that existed independently of human belief and who had actually built the universe. They certainly couldn't have been gods of the sort you got now, who by all accounts were largely incapable of making a cup of coffee.

The universe inside the Project was hurtling through its high-speed time and there was still nothing in there that was even vaguely homely for humans. It was all too hot or too cold or too empty or too crushed. And, distressingly, there was no sign of narrativium.

Admittedly, it has never been isolated on Discworld either, but its existence had long ago been inferred, as the philosopher Lye Tin Wheedle had put it: 'in the same way that milk infers cows'. It might not even have a discrete existence. It might be a particular way in which every other element spun through history, something that they had but did not actually possess, like the gleam on the skin of a polished apple. It was the *glue* of the universe, the frame that held all the others, the thing that told the world what it was going to be, that gave it purpose and direction. You could detect narrativium, in fact, by simply thinking about the world.

Without it, apparently, everything all was just balls spinning in circles, without meaning.

He doodled on the pad in front of him:

There are no turtles anywhere.

'Eat hot plasma! Oh … sorry, sir.'

Ponder peered over his defensive screen.

'When worlds collide, young man, someone is doing something wrong!'

That was the voice of the Senior Wrangler. It sounded more petulant than usual.

Ponder went to see what was going on.

TWELVE
WHERE DO RULES COME FROM?

SOMETHING IS MAKING ROUNDWORLD DO STRANGE THINGS …

It seems to be obeying rules.

Or maybe it's just making them up as it goes along.

Isaac Newton taught us that *our* universe runs on rules, and they are mathematical. In his day they were called 'laws of nature', but 'law' is too strong a word, too final, too arrogant. But it does seem that there are more or less deep patterns in how the universe works. Human beings can formulate those patterns as mathematical rules, and use the resulting descriptions to work out some aspects of nature that would otherwise be totally mysterious, and even exploit them to make tools, vehicles, technology.

Thomas Malthus changed a lot of people's minds when he found a mathematical rule for social behaviour. He said that food grows arithmetically (1-2-3-4-5), but populations grow geometrically (1-2-4-8-16). Whatever the growth rates, eventually population will outstrip food supply: there are limits to growth.* Malthus's law shows that there are rules Down Here as well as Up There, and it tells us that poverty is not the result of evil or sin. Rules can have deep implications.

What are rules? Do they tell us how the universe 'really' works, or do our pattern-seeking brains invent or select them?

There are two main viewpoints here. One is fundamentalist at heart, as fundamentalist as the Taliban and Southern Baptists – indeed, as fundamentalist as the exquisitor Vorbis in *Small Gods* who states his position thus: '… that which appears to our senses is

---

* This rule does require some special assumptions, such as the chronic and irreversible stupidity of humanity.

not the *fundamental* truth. Things that are seen and heard and done by the flesh are mere shadows of a deeper reality.'

Scientific fundamentalism holds that there is *one* set of rules, the Theory of Everything, which doesn't just describe nature rather well, but *is* nature. For about three centuries science seems to have been converging on just such a system: the deeper our theories of nature become, the simpler they become too. The philosophy behind this view is known as reductionism, and it proceeds by taking things to bits, seeing what the bits are and how they fit together, and using the bits to explain the whole. It's a very effective research strategy, and it's served us well for a long time. We've now managed to reduce our deepest theories to just two: quantum mechanics and relativity.

Quantum mechanics set out to describe the universe on very small scales, subatomic scales, but then became involved in the largest scales of all, the origin of the universe in the Big Bang. Relativity set out to describe the universe on very large scales, supergalactic ones, but then became involved in the smallest scales of all, the quantum effects of gravity. Despite this, the two theories disagree in fundamental ways about the nature of the universe and what rules it obeys. The Theory of Everything, it is hoped, will subtly modify both theories in such a way that they fit seamlessly together into a unified whole, while continuing to work well in their respective domains. With everything reduced to one Ultimate Rule, reductionism will have reached the end of its quest, and the universe will be completely explained.

The extreme version of the alternative view is that there are *no* ultimate rules, indeed that there are no totally accurate rules either. What we call laws of nature are human approximations to regularities that crop up in certain specialized regions of the universe – chemical molecules, galaxy dynamics, whatever. There is no reason why our formulations of regularities in molecules and regularities in galaxies should be part of some deeper set of regularities that explains both, any more than chess and soccer should somehow be aspects of the same greater game. The universe could perfectly well be patterned on all levels, without there being an ultimate pattern from which all the others must logically follow. In this view, each set of rules is accompanied by a statement of which areas it can safely be used to describe – 'use these rules for molecules with fewer than a hundred atoms' or 'this rule works for galaxies provided you don't ask about the stars that make them up'. Many such rules are contextual rather than reductionist: they explain why things work the way they do in terms of what is *outside* them.

Evolution, especially before it was interpreted through the eyes of DNA, is one of the clearest examples of this style of reasoning. Animals evolve because of the environment in which they live, including other animals. A curious feature of this viewpoint is that to a great extent the system builds its own rules, as well as obeying them. It is rather like a game of chess played with tiles that can be used to build new bits of board, upon which new kinds of chess piece can move in new ways.

Could the entire universe sometimes build its own rules as it proceeds? We've suggested as much a couple of times: here's a sense in which it might happen. It's hard to see how rules for matter could meaningfully 'exist' when there is no matter, only radiation – as there was at an early stage of the Big Bang. Fundamentalists would maintain that the rules for matter were always implicit in the Theory of Everything, and *became* explicit when matter appeared. We wonder whether the same 'phase transition' that created matter might *also* have created its rules. Physics might not be like that, but biology surely is. Before organisms appeared, there couldn't have been any rules for evolution.

For a more homely example, think of a stone rolling down a bumpy hillside, skidding on a clump of grass, bouncing wildly off bigger rocks, splashing through muddy puddles, and eventually coming to rest against the trunk of a tree. If fundamentalist reductionism is right, then every aspect of the stone's movement, right down to how the blades of grass get crushed, what pattern the mud makes when it splatters, and why the tree is growing where it is anyway, are consequences of one set of rules, that Theory of Everything. The stone 'knows' how to roll, skid, bounce, splash, and stop *because* the Theory of Everything tells it what to do. More than that: because the Theory of Everything is *true*, the stone *itself* is tracking through the logical consequences of those rules as it skitters down the hillside. In principle you could predict that the stone would hit that particular tree, just by working out necessary consequences of the Theory of Everything.

The picture of causality that this viewpoint evokes is one in which the only reasons for things to happen are because the Theory of Everything says so. The alternative is that the universe is doing whatever the universe does, and the stone is in a sense *exploring* the consequences of what the universe does. It doesn't 'know' that it will skid on grass until it hits some grass and finds itself skidding. It doesn't 'know' how to splash mud all over the place, but when it hits the puddle, that's what happens. And so on. Then we humans come along and look at what the stone does, and start finding patterns. 'Yes, the reason it skids is because friction works like *this* …' 'And the laws of fluid dynamics tell us that the mud must scatter like *that* …'

We know that these human-level rules are approximate descriptions, because that's why we invented them. Mud is lumpy, but the rules of fluid dynamics don't take account of lumps. Friction is something rather complicated involving molecules sticking together and pulling apart again, but we can capture a lot of what it does by thinking of it as a force that opposes moving bodies when in contact with surfaces. Because our human-level theories are approximations, we get very excited when some more general principle leads to more accurate results. We then, unless we are careful, confuse 'the new theory gives results that are closer to reality than the old' with 'the new theory's rules are closer to the real rules of the universe than the old one's rules were'. But that doesn't follow: we might be getting a more accurate *description* even though our rules differ from whatever the universe 'really' does. What it really does may not involve following neat, tidy rules at all.

There is a big gap between writing down a Theory of Everything and understanding its consequences. There are mathematical systems that demonstrate this point, and one of the simplest is

Langton's Ant, now the small star of a computer program. The Ant wanders around on an infinite square grid. Every time it comes to a square, the square changes colour from black to white or from white to black, and if it lands on a white square then it turns right, but if it lands on a black square then it turns left. So we know the Theory of Everything for the Ant's universe – the rule that governs its complete behaviour by fixing what can happen on the small scale – and everything that happens in that universe is 'explained' by that rule.

When you set the Ant in motion, what you actually see is three separate modes of behaviour. *Everybody* – mathematician or not – immediately spots them. Something in our minds makes us sensitive to the difference, and it's got nothing to do with the rule. It's the same rule all the time, but we see three distinct phases:

• SIMPLICITY: During the first two or three hundred moves of the Ant, starting on a completely white grid, it creates tiny little patterns which are very simple and often very symmetric. And you sit there thinking 'Of course, we've got a simple rule, so that will give simple *patterns*, and we ought to be able to describe everything that happens in a simple way.'

• CHAOS: Then, suddenly, you notice it's not like that any more. You've got a big irregular patch of black and white squares, and the Ant is wandering around in some sort of random walk, and you can't see any structure at all. For Langton's Ant this kind of pseudo-random motion happens for about the next 10,000 steps. So if your computer is not very fast you can sit there for a long time saying 'Nothing interesting is going to happen, it's going to go on like this forever, it's just random.' No, it's obeying the same rule as before. It's just that to us it *looks* random.

• EMERGENT ORDER: Finally the Ant locks into a particular kind of repetitive behaviour, and it builds a 'highway'. It goes through a cycle of 104 steps, after which it has moved out two squares diagonally and the shape and the colours along the edge are the same as they were at the beginning of that cycle. So that cycle repeats forever, and the Ant just builds a diagonal highway – for ever.

Those three modes of activity are all consequences of the *same* rule, but they are on different levels from the rule itself. There are no rules that talk about highways. The highway is clearly a simple thing, but a 104-step cycle isn't a terribly obvious consequence of the rule. In fact the only way mathematicians can *prove* that the Ant really does build its highway is to track through those 10,000 steps. At that point you could say '*Now* we understand why Langton's Ant builds a highway.' But no sooner.

However, if we ask a slightly more general question, we realize that we don't *understand* Langton's Ant at all. Suppose that before the Ant starts we give it an environment – we paint a few squares black. Now let's ask a simple question: does the Ant always end up building a highway? Nobody knows. All of the experiments on computers suggest that it does. On the other hand, nobody can *prove* that it does. There might be some very strange configuration of squares, and when you start it off on that it gets triggered into some totally different behaviour. Or it could just be a much bigger highway. Perhaps there is a cycle of 1,349,772,115,998 steps that builds a different kind of highway, if only you start from the right thing. We don't know. So for this very simple mathematical system, with one simple rule, and a very simple question, where we *know* the Theory of Everything … it doesn't tell us the answer.

Langton's Ant will be our icon for a very important idea: *emergence*. Simple rules may lead to large, complex patterns. The issue here is not what the universe 'really does'. It is how we understand things and how we structure them in our minds. The simple Ant and its tiled universe are technically a 'complex system' (it consists of a large number of entities that interact with each other, even though most of those entities are simply squares that change colour when an Ant walks on them).

We can create a system, and give it simple rules which 'common sense' suggests should lead to a rather dull future, and we will often find that quite complex features will result. And they will be 'emergent' – that is, we have no practical way of working out what they are going to be apart from … well, watching. The Ant must dance. There are no short cuts.

Emergent phenomena, which you can't predict ahead of time, are just as causal as the non-emergent ones: they *are* logical consequences of the rules. And you have no idea what they are going to be. A computer will not help – all it will do is run the Ant very fast.

A 'geographical' image is useful here. The 'phase space' of a system is the space of all possible states or behaviours – all of the things that the system could do, not just what it *does* do. The phase space of Langton's Ant consists of all possible ways to put black and white squares on a grid – not just the ones that the Ant puts there when it follows its rules. The phase space for evolution is all conceivable organisms, not just the ones that have existed so far. Discworld is one 'point' in the phase space of consistent universes. Phase spaces deal with everything that might be, not what is.

In this imagery, the features of a system are structures in phase space that give it a well-defined 'geography'. The phase space of an emergent system is indescribably complicated: a generic term for such phase spaces is 'Ant Country', which you can think of as a computational form of infinite suburbia. To *understand* an emergent feature you would have to find it *without* traversing Ant Country step by step. The same problem arises when you try to start from a Theory of Everything and work out *what it implies*. You may have pinned down the micro-rules, but that doesn't mean that you understand their macro-consequences. A Theory of Everything would tell you what the *problem* is, in precise language, but that might not help you solve it.

Suppose, for instance, that we had very accurate rules for fundamental particles, rules that really do govern everything about them. Despite that, it's pretty clear that those rules would not greatly help our understanding of something like economics. We want to understand someone who goes into a supermarket, buys some bananas, and pays over some money. How do we approach that from the particle rules? We have to write down an equation for every particle in the customer's body, in the bananas, in the note that passes from customer to cashier. Our description of the transaction – money for bananas – *and* our explanation of it is in terms of an incredibly complicated equation about fundamental particles.

Solving that equation is even harder. *And it might not even be the only fruit they buy.*

We're not saying that the universe hasn't *done* it that way. We're saying that even if it has, that won't help us *understand* anything. So there's a big, emergent gap between the Theory of Everything and its consequences.

A lot of philosophers seem to have got the idea that in an emergent phenomenon the chain of causality is *broken*. If our thoughts are emergent properties of our brain, then to many philosophers they are not physically caused by the nerve cells, the electrical currents, and the chemicals in the brain. We don't mean that. We think it's confused nonsense. We're perfectly happy that our thoughts are *caused* by those physical entities, but you can't describe someone's perceptions or memory in terms of electrical currents and chemicals.

Human beings never understand things that way. They understand things by keeping them simple – in Archchancellor Ridcully's case, the simpler the better. A little narrativium goes a long way: the simpler the story, the better you understand it. Storytelling is the opposite of reductionism; 26 letters and some rules of grammar are no story at all.

One set of modern physical rules poses more philosophical questions than all the others combined: Quantum Mechanics. Newton's rules explained the universe in terms of force, position, speed, and the like – things that make intuitive sense to human beings and let us tell good stories. A century or so ago, however, it became clear that the universe's hidden wiring has other, less intuitive layers. Concepts such as position and speed not only ceased to be fundamental – they ceased to have a well defined meaning at all.

This new layer of explanation, quantum theory, tells us that on small scales the rules are random. Instead of something happening or not, it may do a bit of both. Empty space is a seething mass of potentialities, and time is something you can borrow and pay back again if you do it quickly enough for the universe not to notice. And the Heisenberg Uncertainty Principle says that if you know where something is then you can't also know how fast it's going. Ponder Stibbons would consider himself lucky if he did not have to explain this to his Archchancellor.

A thorough discussion of the quantum world would need a book all to itself, but there's one topic that benefits from some Discworld insights. This is the notorious case of the cat in the box. Quantum objects obey Schrödinger's Equation, a rule named after Erwin Schrödinger which describes how 'wave functions' – waves of quantum existence – propagate through space and time. Atoms and their sub-atomic components aren't really particles: they're quantum wave functions.

The early pioneers of quantum mechanics had enough problems *solving* Schrödinger's equation: they didn't want to worry about what it *meant*. So they spatchcocked together a cop-out clause, the 'Copenhagen interpretation' of quantum observations. This says that whenever you try to observe a quantum wave function it immediately 'collapses' to give a single particle-like answer. This seems to promote the human mind to a special status – it has even been suggested that our purpose in the universe is to observe it, thereby ensuring its existence, an idea that the wizards of UU consider to be simple common sense.

Schrödinger, however, thought this was silly, and in support he introduced a thought experiment now called Schrödinger's Cat. Imagine a box, with a lid that can be sealed so tightly that *nothing*, not even the barest hint of a quantum wavelet, can leak out. The box contains a radioactive atom, which at some random moment will decay and emit a particle, and a particle detector that releases poison gas when it detects the atom decaying. Put the cat in the box and close the lid. Wait a bit.

Is the cat alive or dead?

If the atom has decayed, then the cat's dead. If not, it's alive. However, the box is sealed, so you can't observe what's inside. Since unobserved quantum systems are waves, the quantum rules tell us that the atom must be in a 'mixed' state – half decayed and half not. Therefore the cat, which is a collection of atoms and so can be considered as a gigantic quantum system, is also in a mixed state: half alive, half dead. In 1935 Schrödinger pointed out that cats aren't like that. Cats are macroscopic systems with classical yes/no physics. His point was that the Copenhagen interpretation does not explain – or even address – the link from microscopic quantum physics to macroscopic classical physics. The Copenhagen interpretation replaces a complex physical process (which we don't understand) by a piece of magic: the wave collapses as soon as you try to observe it.

Most of the time this problem is discussed, physicists manage to turn Schrödinger's point on its head. 'No, quantum waves really *are* like that!' And they've done lots of experiments to prove they're right. Except … those experiments have no box, no poison gas, no alive, no dead, and no cat. What they have is quantum-scale analogues – an electron for a cat, positive spin for alive and negative for dead, and a box with Chinese walls, through which anything *can* be observed, but you take great care not to notice.

These discussions and experiments are lies-to-children: their aim is to convince the next generation of physicists that *quantum*-level systems do actually behave in the bizarre way that they do. Fine … but it's got nothing to do with cats. The wizards of Unseen University, who know nothing about electrons but have an intimate familiarity with cats, wouldn't be fooled for an instant. Neither would the witch Gytha Ogg, whose cat Greebo is shut in a box in *Lords and Ladies*. Greebo is the sort of cat that would take on a ferocious wolf and eat it.* In *Witches Abroad* he eats a vampire by accident, and the witches can't understand why the local villagers are so ecstatic.

Greebo has his own way of handling quantum paradoxes. 'Greebo had spent an irritating two minutes in that box. Technically, a cat locked in a box may be alive or it may be dead. You

---

* As Nanny Ogg always says, 'He's just a big softy.'

never know until you look. In fact, the mere act of opening the box will determine the state of the cat, although in this case there were three determinate states the cat could be in: these being Alive, Dead, and Bloody Furious.'

Schrödinger would have applauded. He wasn't talking about quantum states: he wanted to know how they led to ordinary, classical physics in the large, and he could see that the Copenhagen interpretation didn't have anything to say about that. So how *do* classical yes/no answers emerge from quantum Ant Country? The closest we have to an answer is something called 'decoherence', which has been studied by a number of physicists, among them Anthony Leggett, Roland Omnés, Serge Haroche and Luis Davidovich. If you have a big collection of quantum waves and you leave it to its own devices, then the component waves get out of step and fuzz out. This is what a classical object is 'really' like from the quantum standpoint, and it means that cats do, in fact, behave like cats. Experiments show that the same is true even when the role of the detector is played by a microscopic quantum object: a photon's wave function can collapse without any observers being aware, at the time, that it has done so . Even with a quantum cat, death occurs at the instant that the *detector* notices that the atom has decayed. It doesn't require a mind.

In short, Archchancellor, the universe always notices the cat. And a tree in a forest does make a sound when it falls, even if no one is around. The forest is always there.

THIRTEEN

NO, IT CAN'T DO THAT

 ARCHCHANCELLOR RIDCULLY LOOKED AROUND at his colleagues. They'd chosen the long table in the Great Hall for the meeting, since the HEM was getting too crowded.

'All here? Good,' he said. 'Carry on, Mister Stibbons.'

Ponder sifted through his papers.

'I've, er, asked for this meeting,' he said, 'because I'm afraid we're doing things wrong.'

'How can that be?' said the Dean. 'It's *our* universe!'

'Yes, Dean. And, er, no. It's made up its own rules.'

'No, no, it can't do that,' said the Archchancellor. 'We're intelligent creatures. We make the rules. Lumps of rock don't make rules.'

'Not *exactly*, sir,' said Ponder, employing the phrase in its traditional sense of 'absolutely wrong'. 'There are some rules in the Project.'

'How? Is someone else meddling with it?' the Dean demanded. 'Has a Creator turned up?'

'An interesting thought, sir. I'm not qualified to answer that one. The point I'm trying to make is that if we want to do anything constructive, we've got to obey the rules.'

The Lecturer in Recent Runes looked down at the table in front of him. It had been laid for lunch.

'I don't see why,' he said. 'This knife and fork don't tell me how to eat.'

'Er ... in fact, sir, they do. In a roundabout way.'

'Are you trying to tell us that the rules are built in?' said Ridcully.

'Yes, sir. Like: big rocks are heavier than small rocks.'

'That's not a rule, man, that's just common sense!'

'Yes, sir. It's just that the more I look into the Project, the more I'm not sure any more what common sense is. Sir, if we're going to build a world it has to be a ball. A big ball.'

'That's a lot of outmoded religious nonsense, Mister Stibbons.'*

'Yes, sir. But in the Project universe, it's real. Some of the ba … the spheres the students have made are huge.'

'Yes, I've seen them. Showy, to my mind.'

'I was thinking of something smaller, sir. And … and I'm pretty sure things will stay on it. I've been experimenting.'

'Experimenting?' said the Dean. 'What good does *that* do?'

The doors were flung open. Turnipseed, Ponder's assistant, hurried across to the table in a state of some agitation.

'Mister Stibbons! Hex has found something!'

The wizards turned to stare at him. He shrugged.

'It's gold,' he said.

'The Guild of Alchemists is *not* going to be happy about this,' said the Senior Wrangler, as the entire faculty clustered around the project. 'You know what they are for demarcation.'

'Fair enough,' said Ridcully, steering the omniscope. 'We'll just give them a few minutes to turn up, otherwise we'll go on as we are, all right?'

'How can we get it out?' said the Dean.

Ponder looked horrified. 'Sir! This is a universe! It is not a piggy-bank! You can't just turn it upside down, stick a knife in the slot and rattle it around!'

'I don't see why not,' said Ridcully, without looking up. 'It's what people do all the time.' He adjusted the focus. 'Personally I'm *glad* nothing can get out of the thing, though. Call me old fashioned, but I don't intend to occupy the same room as a million miles of exploding gas. What happened?'

'Hex says one of the new stars exploded.'

* Omnianism had taught for thousands of years that the Discworld was in fact a sphere, and violently persecuted those who preferred to believe the evidence of their own eyes. At the time of writing, Omnianism was teaching that there was something to be said for every point of view.

'They're too big to be stars, Ponder. We've been into this.'

'Yes, sir,' Ponder disagreed.

'They've only been around for five minutes.'

'A few days, sir. But millions of years in Project time. People have been dumping rubbish into it, and I think some just drifted in and ... I don't think it was a very well-made st – furnace in the first place.'

The exploding star was shrinking now, but flinging out a great halo of brilliant gases that even lit up one side of the rocky lumps the wizards had been making. Things want to come together and get big, Ponder thought. But when they're big enough, they want to explode. Another law.

'There's lead and copper here, too,' said Ridcully. 'We're in the money now, gentlemen. Except that in this universe there's nothing to spend it on. Even so, it seems we're making progress. You're looking peaky, Mister Stibbons. You ought to get some sleep.'

Progress, thought Ponder. Was that what they were making? But without narrativium, how did anything *know*?

It was day four. Ponder had been awake all night. He wasn't sure, but he thought he'd probably been awake the previous night, too. He may have nodded off for a while, pillowing his head on the growing pile of screwed-up pieces of paper, with the Project winking and twinkling in front of him. If so, he'd dreamed of nothing.

But he'd decided that Progress was what you made it.

After breakfast, the wizards looked at the ball which currently occupied the centre of the omniscope.

'Um, I used iron to start with,' said Ponder. 'Well, mostly iron. There's quite a lot of it about. Some of the ices are really nasty things, and rock by itself just sits there. See this one here?'

A smaller ball of rock hung in space a little way away.

'Yes, very dull,' said the Senior Wrangler. 'Why's it got holes all over it?'

'I'm afraid that when I was dropping rocks on the ball of iron there were a few that went out of control.'

'Could happen to anyone, Stibbons,' said the Archchancellor

generously. 'Did you add gold?'

'Oh yes, sir. And other metals.'

'Gold does give a crust some style, I think. Are these volcanoes?'

'Sort of, sir. They are the, er, acne of young worlds. Only unlike ours, where the rock is melted in the internal magical fields generated in the sub-strata, the magma is kept molten by the heat trapped inside the sphere.'

'Very smoky atmosphere. I can hardly see anything.'

'Yes, sir.'

'Well, *I* don't call it *much* of a world,' said the Dean, sniffing. 'Practically red hot, smoke belching out everywhere …'

'The Dean *does* have a point, young man,' said Ridcully. He was extra kind, just to annoy the Dean. 'It's a brave attempt, but you just seem to have made another ball.'

Ponder coughed. 'I just put this one together for demonstration purposes, sir.' He fiddled with the controls of the omniscope. The scene flickered, and changed. 'Now this,' he said, and there was a twinge of pride in his voice, 'is one I made *earlier*.'

They stared into the lens.

'Well? Just more smoke,' said the Dean.

'Cloud, sir, in fact,' said Ponder.

'Well, we can all make clouds of gas –'

'Er … it's water vapour, sir,' said Ponder.

He reached over and adjusted the omniscope.

The room was filled with the roar of the biggest rainstorm of all time.

By lunchtime it was a world of ice.

'And we were doing so well,' said Ridcully.

'I can't think what went wrong,' said Ponder, wringing his hands. 'We were getting seas!'

'Can't we just warm it up?' said the Senior Wrangler.

Ponder sat down on his chair and put his head in his hands.

'Bound to cool a world down, all that rain,' said the Lecturer in Recent Runes, slowly.

'Very good … er, rocks,' said the Dean. He patted Ponder on the back.

'Poor chap looks a bit down,' hissed the Senior Wrangler to Ridcully. 'I don't think he's been eating properly.'

'You mean ... not chewing right?'

'No eating *enough*, Archchancellor.'

The Dean picked up a piece of paper from Ponder's crowded desk.

'I say, look at these,' he said.

On the paper was written, in Ponder's very neat handwriting:

THE RULES
1 Things fall apart, but centres hold.
2 Everything moves in curves.
3 You get balls.
4 Big balls tell space to bend.
5 There are no turtles anywhere.
6 ... It's so depressing.   —

'Always been a bit of a one for rules, our Ponder,' said the Senior Wrangler.

'Number Six doesn't sound incredibly well formulated,' said Ridcully.

'You don't think he's going a bit bursar, do you?' said the Lecturer in Recent Runes.

'He always thinks everything has to *mean* something,' said Ridcully, who generally took the view that trying to find any deep meaning to events was like trying to find reflections in a mirror: you always succeeded, but you didn't learn anything new.

'I suppose we could simply heat the thing up,' said the Senior Wrangler.

'A sun should be easy,' said Ridcully. 'A big ball of fire should be no problem to a thinking wizard.' He cracked his knuckles. 'Get some of the students to put Mister Stibbons to bed. We'll soon have his little world all warm or my name's not Mustrum Ridcully.'

FOURTEEN
DISC WORLDS

 TO THE WIZARDS OF UNSEEN UNIVERSITY, the heavens include two obviously different types of body: stars, which are tiny pinpricks of light, and the sun, which is a hot ball, not too far away, and passes over the Disc during the day and under it at night. It's taken humanity a while to realize that in our universe it's not like that. Our Sun is a star, and like all stars it's *huge*, so those tiny pinpricks must be a very long way off. Moreover, some of the pinpricks that seem to be stars aren't: they betray themselves by moving differently from the rest. These are the planets, which are a lot closer and a lot smaller, and together with the Earth, Moon, and Sun they form the solar system. Our solar system may *look* like a lot of balls whizzing around in some kind of cosmic game of pool, but that doesn't mean that it started out as balls or rock and ice. It is the outcome of a physical process, and the ingredients that went into that process are not obliged to resemble the result that comes out.

The more we learn about the solar system, the more difficult it is to give a plausible answer to the question: how did it start? It is not the 'answer' part that gets harder – it's the plausibility. As we learn more and more about the solar system, the reality-check that our theories have to pass becomes more and more stringent. This is one reason why scientists have a habit of opening up old questions that everybody assumed were settled long ago, and deciding that they weren't. It doesn't mean that scientists are incompetent: it demonstrates their willingness to contemplate new evidence and re-examine old conclusions in its light. Science certainly does not claim to get things right, but it has a good record of ruling out ways to get things wrong.

What must a theory of the formation of the solar system explain? Principally, of course, the planets – nine of them, dotted rather randomly in space; Mercury, Venus, Earth, Mars, Jupiter, Saturn, Uranus, Neptune, Pluto. It must explain their differences in size. Mercury is a mere 3,032 miles (4,878 km) in diameter, whereas Jupiter is 88,750 miles (142,800 km) in diameter – 29 times as big, 24,000 times the volume, an enormous discrepancy. It must explain their differences in chemical composition: Mercury is made of iron, nickel, and silicate rock; Jupiter is made from hydrogen and helium. It must explain why the planets near the Sun are generally smaller than those further out, with the exception of tiny Pluto, out in the cold and the dark. We don't know a great deal about Pluto, but most of what we do know is strange. For instance, all the other planets lie pretty close to a single plane through the centre of the Sun, but Pluto's orbit is inclined at a noticeable angle. All the other planets have orbits that are pretty close to circles, but Pluto's orbit is much more elongated – to the extent that some of the time it is closer to the Sun than Neptune is.

But that's not all that a theory of the origin of the solar system has to get right. Most planets have smaller bodies in orbit around *them* – our own familiar Moon; Phobos and Deimos, the diminutive twin satellites of Mars; Jupiter's 16 satellites; Saturn's 17 … Even Pluto has a satellite, called Charon, and *that's* weird too. Saturn goes one better and also has entire *rings* of smaller bodies surrounding it, a broad, thin band of encircling rocks that breaks up into a myriad distinct ringlets, with satellites mixed up among them as well as more conventional satellites elsewhere. Then there are the asteroids, thousands of small bodies, some spherical like planets, others irregular lumps of rock, most of which orbit between Mars and Jupiter – except for quite a few that don't. There are comets, which fall in towards the Sun from the huge 'Oort cloud' way out beyond the orbit of Pluto – a cloud that contains *trillions* of comets. There is the Kuiper belt, a bit like the asteroid belt but outside Pluto's orbit: we know over 30 bodies out there now, but we suspect there are hundreds of thousands.

These bodies are known as 'Kuiper Belt Objects' or KBOs. A few years back there was a big fuss because some astronomers wanted to redefine Pluto as a KBO rather than a planet. Pluto probably wouldn't have minded either way, but an awful lot of textbook publishers would have. The scientific case was strong: Pluto is weird in almost every respect, as we've just seen, and it could easily be a KBO that accidentally strayed into the outer reaches of the solar system when disturbed by other bodies. If so, that would explain why it's so weird. It doesn't look like a planet because it isn't one. Other astronomers disagreed strongly with this proposal – for sentimental reasons, for historical ones, or because we don't know for sure that Pluto is a wandering KBO. In the end, Pluto remained on the list of planets. But whether it can hang on to that status for much longer is unclear.

Then there are meteorites, lumps of rock of various sizes that wander erratically through the whole thing ...

Each of these celestial objects, moreover, is a one-off. Mercury is a blisteringly hot lump of cratered rock. Venus has a sulphuric acid atmosphere, rotates the wrong way compared to nearly everything else in the solar system, and is believed to resurface itself every hundred million years or so in a vast, planetwide surge of volcanic activity. Earth has oceans and supports life; since we live on it we find it the most congenial of the planets, but many aliens would probably be aghast at its deadly, poisonous, corrosive oxygen atmosphere. Mars has rock-strewn deserts and dry ice at its poles. Jupiter is a gas giant, with a core of hydrogen compressed so much that it has become metallic, and maybe a small rocky core inside that – 'small' compared to Jupiter, but about three times the diameter of the Earth. Saturn has its rings – but so do Jupiter, Uranus, and Neptune, though these are nowhere near as extensive or spectacular. Uranus has an icy mantle of methane and ammonia, and its axis of rotation is tilted so far that it is slightly upside down. Neptune is similar to Uranus but without that ridiculous axial tilt. Pluto, as we've said, is just crazy. We don't even know accurately how big it is or how massive it is, but it's a Lilliputian in the country of the Gas Giants.

Right ... *all that* is what a theory of the origins of the solar system has to explain. It was all a lot easier when we thought there were six planets, plus the Sun and the Moon, and that was *it*. As for the solar system being an act of special creation by a supernatural being – why would any self-respecting supernatural being make the thing so *complicated*?

Because it makes *itself* complicated – that's why. We now think that the solar system was formed as a complete package, starting from quite complicated ingredients. But it us took a while to realize this.

The first theory of planetary formation that makes any kind of sense by modern standards was thought up by the great German philosopher Immanuel Kant about 250 years ago. Kant envisaged it all starting as a vast cloud of matter – big lumps, small lumps, dust, gas – which attracted each other gravitationally and clumped together.

About 40 years later the French mathematician Pierre-Simon de Laplace came up with an alternative theory of enormous intrinsic beauty, whose sole flaw is that it doesn't actually work. Laplace thought that the Sun formed before the planets did, perhaps by some cosmic aggregation process like Kant's. However, that ancient Sun was much bigger than today's, because it hadn't fully collected together, and the outer fringes of its atmosphere extended well beyond what is now the orbit of Pluto. Like the wizards of Unseen University, Laplace thought of the Sun as a gigantic fire whose fuel must be slowly burning away. As the Sun aged, it would cool down. Cool gas contracts, so the Sun would shrink.

Now comes a neat peculiarity of moving bodies, a consequence of another of Newton's laws, the Law(s) of Motion. Associated with any spinning body is a quantity called 'angular momentum' – a combination of how much mass it contains, how fast it is spinning, and how far out from the centre the spinning takes place. According to Newton, angular momentum is conserved – it can be redistributed, but it neither goes away nor appears of its own accord. If a spinning body contracts, but the rate of spin doesn't change, angular momentum will be lost: therefore the rate of spin must increase to compensate. This is how ice skaters do rapid spins: they start with a slow spin, arms extended, and then bring their arms in close to their body. Moreover, spinning matter experiences a force, centrifugal force, which seems to pull it outwards, away from its centre.

Laplace wondered whether centrifugal force acting on a spinning gascloud might throw off a belt of gas round the equator. He calculated that this ought to happen whenever the gravitational force attracting that belt towards the centre was equal to the centrifugal force trying to fling it away. This process would happen not once, but several times, as the gas continued to contract – so the shrinking Sun would surround itself with a series of rings of material, all lying in the same plane as the Sun's equator. Now suppose that each belt coalesced into a single body ... Planets!

What Laplace's theory got right, but Kant's did not, was that the planets lie roughly in a plane and they all rotate round the Sun in the same direction that the Sun spins. As a bonus, something rather similar might have occurred while those belts were coalescing into planets, in which case the motion of satellites is explained as well. It's not hard to combine the best features of Kant's and Laplace's theories, and this combination satisfied scientists for about a century. However, it slowly became clear that our solar system is far more unruly than either Kant or Laplace had recognized. Asteroids have wild orbits, and some satellites revolve the wrong way. The Sun contains 99% of the solar system's mass, but the planets possess 99% of its angular momentum: either the Sun is rotating too slowly or the planets are revolving too quickly.

As the twentieth century opened, these deficiencies of the Laplacian theory became too great for astronomers to bear, and several people independently came up with the idea that a star developed a solar system when it made a close encounter with another star. As the two stars whizzed past each other, the gravitational attraction from one of them was supposed to draw out a long cigar-shaped blob of matter from the other, which then condensed into planets. The advantage of the cigar shape was that it was thin at the ends and thick at the middle, just as the planets are small close to the Sun or out by Pluto, but big in the middle where Jupiter and Saturn live. Mind you, it was never entirely clear *why* the blob had to be cigar-shaped ...

One important feature of this theory was the implication that solar systems are rather uncommon, because stars are quite thinly scattered and seldom get close enough together to share a mutual cigar. If you were the sort of person who'd be comforted by the idea that human beings are unique in the universe, then this was a rather appealing suggestion: if planets were rare, then *inhabited* planets would be rarer still. If you were the sort of person who preferred to think that the Earth isn't especially unusual, and neither are its life-forms, then the cigar theory definitely put a crimp on the imagination.

By the middle of the twentieth century, the shared-cigar theory had turned out to be even less likely than the Kant-Laplace theory. If you rip a lot of hot gas from the atmosphere of a star, it doesn't condense into planets – it disperses into the unfathomable depths of interstellar space like a drop of ink in a raging ocean. But by then, astronomers were getting a much clearer idea of how *stars* originated, and it was becoming clear that planets must be created by the same processes that produce the stars. A solar system is not a Sun that later acquires some tiny companions: it all comes as one package, right from the start. That package is a disc – the nearest thing in our universe (so far as we know) to Discworld. But the disc begins as a cloud and eventually turns into a lot of balls (Stibbons's Third Rule).

Before the disc formed, the solar system and the Sun started out as a random portion of a cloud of interstellar gas and dust. Random jigglings triggered a collapse of the dustcloud, with everything heading for roughly – but not exactly – the same central point. All it takes to start such a collapse is a concentration of matter somewhere, whose gravity then pulls more matter towards it: random jigglings will produce such a concentration if you wait long enough. Once the process has started, it is surprisingly rapid, taking about ten million years from start to finish. At first the collapsing cloud is roughly spherical. However, it is being carried along by the rotation of the entire galaxy, so its outer edge (relative to the centre of the galaxy) moves more slowly than its inner edge. Conservation of angular momentum tells us that as the cloud collapses it must start spinning, and the more it collapses, the faster it spins. As its rate of spin increases, the cloud flattens out into a rough disc.

More careful calculations show that near the middle this disc thickens out into a dense blob, and most of the matter ends up in the blob. The blob condenses further, its gravitational energy gets traded for heat energy, and its temperature goes up *fast*. When the temperature rises enough, nuclear reactions are ignited: the blob has become a star. While this is happening, the material in the disk undergoes random collisions, just as Kant imagined, and coalesces in a not terribly ordered way. Some clumps get shoved into wildly eccentric orbits, or swung out of the plane of the disc; most clumps, however, are better behaved and turn into decent, sensible planets. A miniature version of the self-same processes can equip most of those planets with satellites.

The chemistry fits, too. Near the Sun, those incipient planets get very hot – too hot for solid water to form. Further out – around the orbit of Jupiter for a dustcloud suitable for making our Sun and solar system – water can freeze into solid ice. This distinction is important for the chemical composition of the planets, and we can see the main outlines if we focus on just three elements: hydrogen, oxygen, and silicon. Hydrogen and oxygen happen to be the two most abundant elements in the universe, apart from helium which doesn't undergo chemical reactions. Silicon is less abundant but still common. When silicon and oxygen combine together, you get silicates – rocks. But even if the oxygen can mop up all the available silicon, some 96% of the oxygen is still unattached, and it combines with hydrogen to make water. There is so much hydrogen – a thousand times as much as oxygen – that virtually all of the oxygen that doesn't go into rocks gets locked away in water. So by far the most common compound in the condensing disc is water.

Close to the star, that water is liquid, even vapour, but out at Jovian distances, it's solid ice. You can pick up a lot of solid mass if you're condensing in a region where ice can form. So the planets there are bigger, and (at least to begin with) they are icy. Nearer the star, the planets are smaller, and rocky. But now the big guys can parlay their initial weight advantage into an even bigger one. Anything that is ten times the mass of the Earth, or greater, can attract *and retain* the two most abundant elements of the disc, hydrogen and helium. So the big balls soak up large amount of extra mass in the form of these two gases. They can also retain compounds like methane and ammonia, which are volatile gases closer to the star.

This theory explains rather a lot. It gets all the main features of the solar system pretty much right. It allows for the odd exceptional motion, but not too many. It agrees with observations of condensing gas clouds in distant regions of space. It may not be perfect, and some special pleading might be necessary to explain odd things like Pluto, but most of the important features click neatly into place.

It also seems likely that huge numbers of planets exist without a central star. In 2000 a team led by Rafael Rebolo observed isolated large planets. A survey of such bodies in the Sigma Orionis cluster shows that the smaller these bodies are, the more numerous they become. If this relationship continues down to Earth-sized bodies (which are too small to observe with current methods) then 'isolated planets' will litter the galaxy. There are probably hundreds of them within 30 light years of Earth, for example. But without a nearby star, there is no way we can observe them directly. There is no star to wobble, no light output to dim as a planet gets in the way, and the planets themselves emit only reflections of distant starlight, far too faint to be seen from here. The conventional theory of planetary formation, in which a star and its accompanying solar system come into being together, cannot apply to such worlds. Small gas clouds are not massive enough to collapse under gravity in the right way, but magnetic effects might cause a collapsing gas cloud round a star to break up and be ejected before its planets are fully formed. Or perhaps these worlds came into being in the usual way, but were then ejected from their solar systems.

The future of the solar system is at least as interesting as its past. The picture of the solar system that emerged from the ideas of Newton and his contemporaries was very much that of a clockwork universe – a celestial machine that, once set ticking, would continue to follow some simple mathematical rules and continue ticking merrily away forever. They even *built* celestial machines, called orreries, with lots and lots of cogwheels, in which little brass planets with ivory moons went round and round when you turned a handle.

We now know that the cosmic clockwork can go haywire. It won't happen quickly, but there may be some big changes to the solar system on the way. The underlying reason is chaos – chaos in the sense of 'chaos theory', with all those fancy multicoloured 'fractal' things, a rapidly expanding area of mathematics which is invading all of the other sciences. Chaos teaches us that simple rules need not lead to simple behaviour – something that Ponder Stibbons and the other wizards are in the process of discovering. In fact, simple rules can lead to behaviour that in certain respects has distinct elements of randomness. Chaotic systems start out behaving predictably, but after you cross some 'prediction horizon' all predictions fail. Weather is chaotic, with a prediction horizon of about four days. The solar system, we now know, is chaotic, with a prediction horizon of tens of millions of years. For example, we can't be sure which side of the Sun Pluto will be in a hundred million years' time. It will be in the same *orbit*, but its position in that orbit is completely uncertain.

We know this because of some mathematical work that was done, in part, with an orrery – but this was a 'digital orrery', a custombuilt computer that could do celestial mechanics very fast. The digital orrery was developed by Jack Wisdom's research group, which – in competition with its rival headed by Jaques Laskar – has been extending our knowledge of the solar system's future. Even though a chaotic system is unpredictable in the long run, you can make a whole series of independent attempts at predicting it and then see what they agree about. According to the mathematics, you can be pretty sure *those* things are right.

One of the most striking results is that the solar system is due to lose a planet. About a billion years from now, Mercury will move outwards from the Sun until it crosses the orbit of Venus. At that point, a close encounter between Venus and Mercury will fling one or the other, possibly both, out of the solar system altogether – unless they hit something on the way, which is highly unlikely, but possible. It might even be the Earth, or the passing Venus might join with us in a cosmic dance whose end result is the *Earth* being flung out of the solar system. The details are unpredictable, but the general scenario is very likely.

This means that we've got the wrong picture of the solar system. On a human timescale it's a very simple place, in which nothing much changes. On its *own* timescale, hundreds of millions of years, it's full of drama and excitement, with planets roaring all over the place, whirling around each other, and dragging each other out of orbit in a mad gravitational dance.

This is vaguely reminiscent of *Worlds in Collision*, a book published in 1950 by Immanuel Velikovsky, who believed that a giant comet was once spat out by Jupiter, passed close to the Earth *twice*, had a love affair with Mars (giving rise to a brood of baby comets), and finally retired to live in peace as Venus. Along the way it gave rise to many strange effects that became stories in the Bible. Velikovsky was right about one thing: the orbits of the planets are not fixed forever. He wasn't right about much else.

Do other solar systems encircle distant stars, or are we unique? Until a few years ago there was a lot of argument about this question, but no hard evidence. Most scientists, if they had to bet, would have backed the existence of other solar systems, because the collapsing dustcloud mechanism could easily get going almost anywhere there's cosmic dust – and there are a hundred billion stars in our own galaxy, let alone the billions upon billions of others in the universe, all of which once *were* cosmic dust. But that's only indirect evidence. Now the position is much clearer. Characteristically, however, the story involves at least one false start, and a critical re-examination of evidence that at first looked rather convincing.

In 1967 Jocelyn Bell, a graduate student at the University of Cambridge, was working for a doctorate under the direction of Anthony Hewish. Their field was radio astronomy. Like light, radio is an electromagnetic wave, and like light, radio waves can be emitted by stars. Those radio waves can be detected using parabolic dish receivers – today's satellite TV dishes are a close relative – rather misleadingly called 'radio telescopes', even though they work on very different principles from normal optical telescopes. If we look at the sky in the radio part of the electromagnetic spectrum, we can often 'see' things that are not apparent using ordinary visible light. This should be no surprise: for example military snipers can 'see in the dark' using infra-red waves – detecting things by the heat they emit. The technology in those days wasn't terribly slick, and the radio signals were recorded on long rolls of paper using automatic pens that drew wiggly curves in good old-fashioned ink. Bell was given the task of looking for interesting things on the paper charts – carefully scanning about 400 feet of chart per week. What she found was very strange – a signal that pulsated about thirty times per second. Hewish was sceptical, suspecting that the signal was somehow generated by their measuring instruments, but Bell was convinced it was genuine. She searched through three miles of previous charts and found several earlier instances of the same signal, which proved she was right. Something out there was emitting the radio equivalent of a reverberating whistle. The object responsible was named a 'pulsar' – a pulsating starlike object.

What could these strange things be? Some people suggested they were radio signals from an alien civilization, but all attempts to extract the alien equivalent of *The Jerry Springer Show* failed (which was possibly just as well). There seemed to be no structured messages hidden in the signals. In fact, what they are now believed to be is even stranger than an alien TV programme. Pulsars are thought to be neutron stars – stars composed of highly degenerate matter containing only neutrons, usually a mere 12 miles (20 km) in diameter.

Recall that neutron stars are incredibly dense, formed when a larger star undergoes gravitational collapse. That initial star, as we have seen, will be spinning, and because of conservation of angular momentum, the resulting neutron star has to spin a lot faster. In fact, it typically spins through about thirty complete revolutions every second. For a star, that's pretty speedy. Only a tiny star like a neutron star can do it: if an ordinary star were to revolve that fast, its surface would have to be travelling faster than light, which wouldn't greatly please Einstein. (More realistically, a normal star would be torn apart at much lower speeds.) But a neutron star is small, and its angular momentum is comparatively large, and pirouetting thirty times a second is no problem at all.

For a helpful analogy, contemplate our own Earth. Like a pulsar, it spins on an axis. Like a pulsar, it has a magnetic field. The magnetic field has an axis too, but it's different from the axis of rotation – that's why magnetic north is not the same as true north. There's no good reason for magnetic north to be the same as true north on a pulsar, either. And if it isn't, that magnetic axis whips round thirty times every second. A rapidly spinning magnetic field emits radiation, known as synchrotron radiation – and it emits it in two narrow beams which point along the magnetic axis. In short, a neutron star projects twin radio beams like the spinning gadgetry on top of a terrestrial lighthouse. So if you look at a neutron star in radio light, you see a bright flash as the beam points towards you, and then virtually nothing until the beam comes round again. Every second, you see thirty flashes. That's what Bell had noticed.

If you're a living creature of remotely orthodox construction, you definitely do not want your star to be a pulsar. Synchrotron radiation is spread over a wide range of wavelengths, from visible light to x-rays, and x-rays can seriously damage the health of any creature of remotely orthodox construction. But no astronomer ever seriously suspected that pulsars might have planets, anyway. If a big star collapses down to an incredibly dense neutron star, surely it will gobble up all the odd bits of matter hanging around nearby. Won't it?

Perhaps not. In 1991 Matthew Bailes announced that he had detected a planet circling the pulsar PSR 1829-10, with the same mass as Uranus, and lying at a distance similar to that of Venus from the Sun. The known pulsars are much too far away for us to see planets directly – indeed all stars, even the nearest ones, are too far away for us to see planets directly. However, you can spot a star that has planets by watching it wiggle as it walks. Stars don't sit motionless in space – they generally seem to be heading somewhere, presumably as the result of the gravitational attraction of the rest of the universe, which is lumpy enough to pull different stars in different directions. Most stars move, near enough, in straight lines. A star with planets, though, is like someone with a dancing partner. As the planets whirl round the star, the star wobbles from side to side. That makes its path across the sky slightly wiggly. Now, if a big fat dancer whirls a tiny feather of a partner around, the fat one hardly moves at all, but if the two partners have equal weight, they both revolve round a common centre. By observing the shape of the wiggles, you can estimate how massive any encircling planets are, and how close to the star their orbits are.

This technique first earned its keep with the discovery of double stars, where the dancing partner is a second star, and the wobbles are fairly pronounced because stars are far more massive than planets. As instrumentation has become more accurate, ever tinier wobbles can be detected, hence ever tinier dancing partners. Bailes announced that pulsar PSR 1829-10 had a dancing partner whose mass was that of a planet. He couldn't observe the wiggles directly, but he could observe the slight changes they produced in the timing of the pulses in the signal. The only puzzling feature was the rotational period of the planet: *exactly* six Earth months. Bit of a coincidence. It quickly turned out that the supposed wiggles were not caused by a planet going round the pulsar, but by a planet much closer to home – Earth. The instruments were doing the wiggling at *this* end, not the pulsar at the far end.

Scarcely had this startling claim of a pulsar planet been withdrawn, however, when Aleksander Wolszczan and Dale Frail announced the discovery of two more planets, both circling pulsar PSR 1257+12. A pulsar solar system with at least two worlds! The way you wiggle when you have two dancing partners is more complex than the way you do it with one, and it's difficult to mistake such a signal for something generated at the receiving end by the motion of the Earth. So this second discovery seems to be fairly solid, unless there is a way for pulsars to vary their output signals in just such a complex manner without having planets – maybe the radio beam could be a bit wobbly? We can't go there to find out, so we have to do the best we can from here; and from here it looks good.

So there do exist planets outside our solar system. But it's the possibility of life that really makes distant planets interesting, and a pulsar planet with all those x-rays is definitely not a place for anything that wants to be alive for very long. But now conventional stars are turning out to have planets, too. In October 1995 Michel Mayor and Didier Queloz found wobbles in the motion of the star 51 Pegasi that were consistent with a planet of about half Jupiter's mass. Their observations were confirmed by Geoffrey Marcy and Paul Butler, who found evidence for two more planets – one seven times the mass of Jupiter orbiting 70 Virginis, and one two or three times Jupiter's mass orbiting 47 Ursae Majoris.

By 1996 seven such planets had been found. As we write, about 70 extrasolar planets have been detected, either by the wobble method, or by observing the light output from a star and seeing whether it changes as an orbiting planet reflects different amounts of its light. Theoretical calculations show that with improved telescopes, this method might even be able to detect how fast the planet is rotating. Even now, new extrasolar planets are being found virtually every week. The exact number fluctuates because every so often astronomers discover problems with previous measurements that cast doubt on somebody else's favourite new planet, but the general trend is up. And our nearest sunlike neighbour, epsilon Eridani, is now known to possess an encircling dustcloud, perhaps like our Sun's Oort cloud, thanks to observations made in 1998 by James Greaves and colleagues. We can't see any wobbles, though, so if it has planets, their mass must be less than three times that of Jupiter. A year earlier, David Trilling and Robert Brown used observations of a similar dustcloud round 55 Cancri, which does wobble, to show that it has a planet whose mass is at most 1.9 Jupiters. This definitely rules out alternative explanations of the unseen companion, for example that it might be a 'brown dwarf' – a failed star.

Although today's telescopes cannot detect an alien planet *directly*, future telescopes might. Conventional astronomical telescopes use a big, slightly dish-shaped mirror to focus incoming light, plus lenses and prisms to pick up the image and send it to what used to be an eyepiece for an astronomer to look down, but then became a photographic plate, and is now likely to be a 'charge-coupled device' – a sensitive electronic light-detector – hooked up to a computer. A single telescope of conventional design would need a very big mirror indeed to spot a planet round another star – a mirror some 100 yards (100 m) across. The biggest mirror in existence today is one-tenth that size, and to see any detail on the alien world you'd need an even bigger mirror, so none of this is really practicable.

But you don't have to use just one telescope.

A technique known as 'interferometry' makes it possible, in principle, to replace a single mirror 100 yards wide by two much smaller mirrors 100 yards apart. Both produce images of the same star or planet, and the incoming light waves that form those images are aligned *very* accurately and combined. The two-mirror system gathers less light than a complete 100-yard mirror would, but it can resolve the same amount of tiny detail. And with modern electronics, very small quantities of incoming light can be amplified. In any case, what you actually do is use dozens of smaller mirrors, together with a lot of clever trickery that keeps them aligned with each other and combines the images that they receive in an effective manner.

Radio astronomers use this technique all the time. The biggest technical problem is keeping the length of the path from the star to its image the same for all of the smaller telescopes, to within an accuracy of one wavelength. The technique is relatively new in optical astronomy, because the wavelength of visible light is far shorter than that of radio waves, but for visible light the real killer is that it's not worth bothering if your telescopes are on the ground. The Earth's atmosphere is in continual turbulent motion, bending incoming light in unpredictable ways. Even a very powerful ground-based telescope will produce a fuzzy image, which is why the Hubble Space Telescope is in orbit round the Earth. Its planned successor, the Next Generation Space Telescope, will be a million miles away, orbiting the Sun, delicately poised at a place called Lagrange point L2. This is a point on the line from the Sun to the Earth, but further out, where the Sun's gravity, the Earth's gravity, and the centrifugal force acting on the orbiting telescope all cancel out. Hubble's structure includes a heavy tube which keeps out unwanted light – especially light reflected from our own planet. It's a lot darker out near L2, and that cumbersome tube can be dispensed with, saving launch fuel. In addition, L2 is a lot colder than low Earth orbit, and that makes infra-red telescopy much more effective.

Interferometry uses a widely separated array of small telescopes instead of one big one, and for optical astronomy the array has to be set up in space. This produces an added advantage, because space is *big* – or, in more Discworldly terms, a place to be big *in*. The biggest distance between telescopes in the array is called the baseline. Out in space you can create interferometers with gigantic baselines – radio astronomers have already made one that is bigger than the Earth by using one ground-based telescope antenna and one in orbit. Both NASA and the European Space Agency ESA have missions on the drawing-board for putting prototype optical interferometer arrays – 'flocks' is a more evocative term – into space.

Some time around 2003, NASA will launch Space Technology 3 (previously named Deep Space 3), involving two spacecraft flying 0.6 miles (1 km) apart and maintaining station relative to each other to a precision of less than half an inch (1 cm). A successor, Star Light, will follow in 2005. Another NASA venture, the Space Interferometry Mission, will employ three interferometers with a 10-metre baseline and is tentatively due to launch in 2009. And NASA is thinking about a Terrestrial Planet Finder in 2012, which will look not just for planets, but for carbon dioxide, water vapour, ozone, and methane, which could be signs of life – or, at least, of a planet that might be able to support life similar to ours. Life Finder, with no specific date, would look more closely.

The European Space Agency (ESA) has similar missions on the drawing-board. SMART-2, consisting of two satellites orbiting in formation, is planned for 2006. A more ambitious ESA project is Darwin, a flotilla of 6 telescopes that could be in space by 2014.

The biggest dream of all, though, is NASA's Planet Imager, pencilled in for 2020. A squadron of five spacecraft, each equipped with four optical telescopes, will deploy itself into an interferometer with a baseline of several thousand miles, and start mapping alien planets. The nearest star is just over four light years away; computer simulations show that 50 telescopes with a baseline of just 95 miles (150 km) can produce images of a planet 10 light years away that are good enough to spot continents and even moons the size of ours. With 150 telescopes and the same baseline, you could look at the Earth from 10 light years away and see hurricanes in its atmosphere. Think what could be done with a thousand-mile baseline.

Planets outside our solar system do exist, then – and they probably exist in abundance. That's good news if you're hoping that somewhere out there are alien lifeforms. The evidence for those, though, is controversial.

Mars, of course, is the traditional place where we expect to find life in the solar system – partly because of myths about Martian 'canals' which astronomers thought they'd seen in their telescopes but which turned out to be illusions when we sent spacecraft out there to take a close look, partly because conditions on Mars are in some ways similar to those on Earth, though generally nastier, and partly because dozens of science-fiction books have subliminally prepared us for the existence of Martians. Life does show up in nasty places here, finding a foothold in volcanic vents, in deserts, and deep in the Earth's rocks. Nevertheless, we've found no signs of life on Mars.

Yet.

For a while, some scientists thought we had. In 1996 NASA announced signs of life on Mars. A meteorite dug up in the Antarctic with the code number ALH84001 had been knocked off Mars 15 million years ago by a collision with an asteroid, and plunged to Earth 13,000 years ago. When it was sliced open and the interior examined at high magnification we found three possible signs of life. These were markings like tiny fossil bacteria, crystals containing iron like those made by certain bacteria, and organic molecules resembling some found in fossil bacteria on Earth. It all pointed to: Martian bacteria! Not surprisingly, this claim led to a big argument, and the upshot is that all three discoveries are almost certainly *not* evidence for life at all. The fossil 'bacteria' are much too small and most of them are steps on crystal surfaces that have caused funny shapes to form in the metal coatings used in electron microscopy; the iron-bearing crystals can be explained without invoking bacteria at all; and the organic molecules could have got there without the aid of Martian life.

However, in 1998 the Mars Global Surveyor did find signs of an ancient ocean on Mars. At some point in the planet's history, huge amounts of water gushed out of the highlands and flowed into the northern lowlands. It was thought that this water just seeped away or evaporated, but it now turns out that the edges of the northern lowlands are all at much the same height – like shorelines eroded by an ocean. The ocean, if it existed, covered a quarter of Mars's surface. If it contained life, there ought to be Martian fossils for us to find, dating from that period.

The current favourite for life in the solar system is a surprise, at least to people who don't read science fiction: Jupiter's satellite Europa. It's a surprise because Europa is exceedingly cold, and covered in thick layers of ice. However, that's not where the life is suspected to live. Europa is held in Jupiter's massive gravitational grasp, and tidal forces warm its interior. This *could* mean that the deeper layers of the ice have melted to form a vast underground ocean. Until recently this was pure conjecture, but the evidence for liquid water beneath Europa's surface has now become very strong indeed. It includes the surface geology, gravitational measurements, and the discovery that Europa's interior conducts electricity. This finding, made in 1998 by K.K.Khurana and others, came from observations of the worldlet's magnetic field made by the space probe Galileo. The shape of the magnetic field is unusual, and the only reasonable explanation so far is the existence of an underground ocean whose dissolved salts make it a weak conductor of electricity. Callisto, another of Jupiter's moons, has a similar magnetic field, and is now also thought to have an underground ocean. In the same year, T.B.McCord and others observed huge patches of hydrated salts (salts whose molecules contain water) on Europa's surface. This might perhaps be a salty crust deposited by upwelling water from a salty ocean.

There are tentative plans to send out a probe to Europa, land it, and drill down to see what's there. The technical problems are formidable – the ice layer is at least ten miles (16 km) thick, and the operation would have to be carried out *very* carefully so as not to disturb or destroy the very thing we're hoping to find: Europan organisms. Less invasively, it would be possible to look for tell-tale molecules of life in Europa's thin atmosphere, and plans are afoot to do this too. Nobody expects to find Europan antelopes, or even fishes, but it would be surprising if Europa's water-based chemistry, apparently an ocean a hundred miles (160 km) deep, has *not* produced life. Almost certainly there are sub-oceanic 'volcanoes' where *very* hot sulphurous water is vented through the ocean floor. These provide a marvellous opportunity for complicated chemistry, much like the chemistry that started life on Earth.

The least controversial possibility would be an array of simple bacteria-like chemical systems forming towers around the hot vents – much as Earthly bacteria do in the Baltic sea. More complicated creatures like amoebas and parameciums would be a pleasant surprise; anything beyond that, such as multicellular organisms, would be a bonus. Don't expect plants – there's not enough light that far from the sun, even if it could filter down through the layers of ice. Europan life would have to be powered by chemical energy, as it is around Earth's underwater volcanic vents. Don't expect Europan lifeforms to look like the ones round *our* vents, though: they will have evolved in a different chemical environment.

In 2001 Brad Dalton, an astrogeophysicist ('geologist' for other worlds) wondered whether we might already have seen alien life. Europa's surface is covered with red-brown marks, especially along what appear to be cracks in the ice. Dalton discovered that the infrared spectrum of these marks is very similar to that of Earthly bacteria that can survive in very cold environments. In fact, for three species of bacteria the spectrum is closer than that for the most likely alternative explanation of the marks: mineral salts seeping up through the cracks. Europa's surface is too cold for even these bacteria, but they might thrive in the ocean and somehow be transported to the surface.

FIFTEEN
THE DAWN
OF DAWN

 PONDER OPENED HIS EYES and looked up into a face out of time. A mug of tea was thrust towards him.

It had a banana stuck in it.

'Ah … Librarian,' said Ponder weakly, taking the cup. He drank, stabbing himself harmlessly in the left eye. The Librarian thought that practically everything could be improved by the addition of soft fruit, but apart from that he was a kindly soul, always ready with a helping hand and a banana.*

The wizards had put Ponder to sleep on a bench in the store-room. Dusty items of magical gear were stacked from floor to ceiling. Most of it was broken, and all of it was covered in dust.

Ponder sat up and yawned.

'What time is it?'

'Ook.'

'Gosh, that late?'

As the warm clouds of sleep ebbed, it dawned on Ponder that he had left the Project entirely in the hands of the senior faculty. The Librarian was impressed at how long the door kept swinging.

Most of the main laboratory was empty, except for the pool of light around the Project.

The Dean's voice said, 'Mappin Winterley … that's a nice name?'

'Shutup.'

'Owen Houseworthy?'

'Shutup.'

---

* A magical accident had once turned the University's Librarian into an orangutan, a state which he enjoyed sufficiently to threaten, with simple and graphic gestures, anyone who suggested turning him back. The wizards noticed no difference now. An orangutan seemed such a *natural* shape for a librarian.

'William.'

'Shut *up*, Dean. That's not funny. It never *was* funny.' This was the voice of the Archchancellor.

'Just as you say, Gertrude.'

Ponder advanced towards the glowing Project.

'Ah, Ponder,' said the Senior Wrangler, stepping in front of it hurriedly. 'Good to see you looking so –'

'You've been … doing things, haven't you,' said Ponder, trying to see around him.

'I'm sure everything can be mended,' said the Lecturer in Recent Runes.

'And it's still *nearly* circular,' said the Dean, 'Just ask Charlie Grinder here. His name's definitely not Mustrum Ridcully, I know that.'

'I'm *warning* you, Dean –'

'What have you *done*?'

Ponder looked at his globe. It was certainly warmer now, and also rather less globular. There were livid red wounds across one side, and the other hemisphere was mainly one big fiery crater. It was spinning gently, wobbling as it did so.

'We've saved most of the bits,' said the Senior Wrangler, watching him hopefully.

'What did you *do*?'

'We were only trying to be helpful,' said the Dean. 'Gertrude here suggested we make a sun, and –'

'Dean?' said Ridcully.

'Yes, Archchancellor?'

'I would just like to point out, Dean, that it was not a very funny joke to begin with. It was a pathetic attempt, Dean, at dragging a sad laugh out of a simple figure of speech. Only four-year-olds and people with a serious humour deficiency keep *on* and *on* about it. I just wanted to bring this out into the open, Dean, calmly and in a spirit of reconciliation, for your own good, in the hope that you may be made well. We are all here for you, although I can't imagine what *you* are here for.' Ridcully turned to the horrified Ponder. 'We made a sun –'

'– some suns –' muttered the Dean.

'– some suns, yes, but ... well, this "falling in circles" business is very difficult, isn't it? Very hard to get the hang of.'

'You crashed a sun into my world?' said Ponder.

'Some suns,' said Ridcully.

'*Mine* bounced off,' said the Dean.

'And created this rather embarrassingly large hole here,' said the Archchancellor. 'And incidentally knocked a huge lump out of the place.'

'But at least *bits* of my sun *burned* for a long time,' said the Dean.

'Yes, but *inside* the world. That doesn't count.' Ridcully sighed. 'Yet your machine, Mister Stibbons, says a sun sixty miles across won't work. And that's ridiculous.'

Ponder stared hollow-eyed at his world, wobbling around like a crippled duck.

'There's no narrativium,' he said dully. 'It doesn't know what size a sun should be.'

'Ook,' said the Librarian.

'Oh dear,' said Ridcully. 'Who let him in here?'

The Librarian was informally banned from the High Energy Magic building, owing to his inherent tendency to check on what things were by tasting them. This worked very well in the Library, where taste had become a precision reference system, but was less useful in a room occasionally containing bus bars throbbing with several thousand thaums. The ban was informal, of course, because anyone capable of pulling the dooknob *right through* an oak door can obviously go where he likes.

The orangutan knuckled over to the dome and tasted it. The wizards tensed as delicate black fingers twiddled the knobs of the omniscope, bringing into focus the furnace that had exploded yesterday. It was a tiny point of light now, surrounded by coruscating streamers of glowing gas.

The focus moved in to the glowing ember.

'Still too big,' said Ridcully. 'Nice try, old chap.'

The Librarian turned towards him, the light of the explosion moving across his face, and Ponder held his breath.

It came out in a rush.

'Someone give me a light!'

The globes on his desk rolled off and bounced on the floor as he tried to grab one. He held it as the Senior Wrangler obligingly lit a match, and waggled it this way and that.

'It'll work!'

'Jolly good!' said Ridcully. 'What will?'

'Days and nights!' said Ponder. 'Seasons, too, if we do it right! Well done, sir! I'm not sure about the wobble, but you might have got it just right!'

'That's the kind of thing we do,' said Ridcully, beaming. 'We're the chaps for getting things right, sure enough. What things did we get right this time?'

'The spin!'

'That was *my* sun that did that,' the Dean pointed out, smugly.

Ponder was almost dancing. And then, suddenly, he looked grave.

'But it all depends on fooling people down there,' he said. 'And there isn't anyone down there ... HEX?'

There was a mechanical rattle as HEX paid attention.

+++ Yes? +++

'Is there any way we can get on to the world?'

+++ Nothing Physical May Enter The Project +++

'I want someone down there to observe things from the surface.'

+++ That Is Possible. Virtually Possible +++

'Virtually?'

+++ But You Will Need A Volunteer. Someone To Fool +++

'This is Unseen University,' said the Archchancellor. 'That should present *no* problem.'

SIXTEEN

EARTH
AND FIRE

 WE DON'T KNOW IF THE EARTH IS A TYPICAL PLANET. We don't know how common 'aqueous' planets with oceans and continents and atmospheres are. In our solar system, Earth is the only one. And we'd better be careful about phrases like 'earthlike planet', because for about half of Earth's history it has not been the familiar blue-green planet that we see in satellite photos, with its oxygen atmosphere, white clouds, and everything else that we are used to. In order to *get* an earthlike planet, in today's sense, you have to start with an unearthlike planet and wait a few billion years. And what you get is quite different from what, only a few decades ago, we *thought* the Earth was like.

We thought it was a very stable place – that if you could go back to the time when the oceans and continents first separated out, they'd have been in the same places they are now. And we thought that the interior of the Earth was pretty simple.

We were wrong.

We know a lot about the surface of the Earth, but we still know much less about what's inside it. We can study the surface by going there, which is usually fairly easy, unless we want to look at the top of Everest. We can also penetrate the ocean depths using vehicles that can protect frail humans against the huge pressures of the deep seas, and we can dig holes down into the ground and send people down those too. We can get further information about the top few miles of the Earth's crust by drilling, but that's just a thin skin, comparatively speaking. We have to infer what it's like deeper down from indirect observations, of which the most important are shock-waves emitted by earthquakes, laboratory experiments, and theory.

The surface of our planet generally seems fairly placid – apart from weather and the sometimes severe effects of the seasons – but there are plenty of volcanoes and earthquakes to remind us that not so far below our feet it's a lot less hospitable. Volcanoes form where the molten rocks inside the Earth well up to the surface, often accompanied by massive clouds of gas or ash, all of it emerging under high pressure. In 1980 Mount St Helens in Washington State, USA blew up like a pressure-cooker whose lid had been tied down, and about half of a large mountain simply disappeared. Earthquakes happen when the Earth's crustal rocks slide past each other along deep cracks. Later we'll see what drives these two things, but they need to be put into perspective: despite occasional disasters, the surface of the Earth has been sufficiently hospitable for life to have evolved and survived for several billion years.

The Earth is nearly spherical, having a diameter of 7,928 miles (12,756 km) at the equator but only 7,902 miles (12,714 km) from pole to pole. The slight broadening at the equator is the result of centrifugal forces from the Earth's spin, and originally set in when the planet was molten. The Earth is the densest planet in the solar system, with an average density 5.5 times that of water. When the Earth condensed from the primal dustcloud the chemical elements and compounds that formed it separated into layers: the denser materials sank to the centre of the Earth and the lighter ones floated to the top, much as a layer of light oil floats on denser water.

In 1952 the American geophysicist Francis Birch set out a description of the general structure of our planet which has been modified in only minor ways since. The inside of the Earth is hot, but the pressure there is also very high: the most extreme conditions occur at the centre where the temperature is about 6,000°C and the pressure is 3 million times atmospheric pressure. Heat tends to make rocks and metals melt, but pressure tends to solidify them, so it is the combination of these two conflicting factors that determines whether the materials are liquid or solid. The centre of the Earth is a rather lumpy spherical core, mainly made of iron, with a radius of roughly 2,220 miles (3,500 km). The innermost regions of the core, out to a radius of 600 miles (1000 km), are solid, but a thick outer layer is molten. The very top layers of the Earth form a thin skin, the crust, which is only a few miles thick. Between crust and core lies the mantle, which is solid, formed from a variety of silicate rocks. The mantle also divides into an inner layer and an outer layer, with the division occurring at a radius of about 3,600 miles (5,800 km). Above this 'transition zone' the main rocks are olivine, pyroxine, and garnet; below it their crystal structures become more tightly packed, forming such minerals as perovskite. The outermost parts of the mantle, and the deeper parts of the crust where the two join, are again molten.

The crust is between 3 and 12 miles (5 and 20 km) thick, and there's a lot going on there. Those parts of the crust that form the continental land masses are mainly made of granite. Beneath the oceans, the crustal layer is predominately basalt, and this basalt layer continues underneath the continental granite. So the continents are broad, thin sheets of granite stuck on top of a basalt skin. From the Earth's surface the most evident features of the granite layers are mountains. The highest ones look big to us, but they rise no more than 5 miles (9 km) above sea level, a mere seventh of a per cent of the Earth's radius. The deepest part of the ocean, the Mariana Trench in the northwest Pacific, plunges 7 miles (11 km) beneath the waves. The overall deviation from an ideal sphere (strictly, spheroid, because of the flattening of the poles) is about one-third of a per cent – about as irregular as the shallow indentations you find on a basketball, which add to its grip. Our home planet, give or take a bit of squashing, is remarkably round and surprisingly smooth. Gravity made it that way, and it keeps it that way – except that some small but interesting movements in the mantle and the crust add a few wrinkles.

How do we know all this? Mainly because of earthquakes. When an earthquake hits, the whole Earth rings like a bell hit by a hammer. Shockwaves, vibrations emitted by the earthquake, travel through the Earth. They are deflected by transition zones between different kinds of material, such as that between core and mantle, or lower and upper mantle. They bounce off the Earth's crust and head back down again. There are several kinds of wave, and they travel with different speeds. So the short sharp shock of an earthquake gives rise to a very complex pattern of waves. When the waves hit the surface they can be detected and recorded, and recordings made in different places can be compared. Working backwards from these recorded signals, it is possible to deduce a certain amount about the underground geography of our planet.

One consequence of the Earth's internal structure is a magnetic field. A compass needle points roughly north. The standard 'lie-to-children' is that the Earth is a giant magnet. Let's unpack the next layer of explanation.

The Earth's magnetic field has long been something of a puzzle since magnets are seldom made out of rock, but once you realize that the Earth has a whopping great lump of iron inside it, everything makes much more sense. The iron doesn't form a 'permanent' magnet, like the ones you inexplicably buy to stick plastic pigs and teddy bears on the fridge door; it's more like a dynamo. In fact it's called the geomagnetic dynamo. The iron in the core is, as we've said, mostly molten, except for a slightly lumpy solid bit in the middle. The liquid part is still heating up – the old explanation of this was that radioactive elements are denser than most of the rest of the Earth, and therefore sank to the middle where they became trapped, and their radioactive energy is showing up as heat. The current theory is quite different: the molten part of the core is heating up because the solid part is cooling down. The liquid iron that is in contact with the solid core is itself slowly solidifying, and when it does so it loses heat. That heat has to go somewhere, and it can't just waft away unnoticed as warm air because everything is thousands of miles underground. So it goes into the molten part of the core and heats it up.

You're probably wondering how the part that is in contact with the solid core can simultaneously be getting cooler, so that it solidifies too, and be getting hotter as a result of that solidification, but what happens is that the hot iron moves away as soon as it's been warmed up. For an analogy, think about a hot air balloon. When you heat air, it rises: the reason is that air expands when it gets hot, so becomes less dense, and less dense things float on top of denser things. A balloon traps the hot air in a huge cloth bag, usually brightly coloured and emblazoned with adverts for banks and estate agents, and floats up along with the air. Now hot iron rises, just as hot air does, and that takes the newly heated iron away from the solid core. It heads upwards, cooling slowly as it does so, and when it gets to the top it cools down – comparatively speaking – and starts to sink again. The result is that the Earth's core circulates up and down, being heated at the bottom and cooling at the top. It can't all go up at the same time, so in some regions it's heading up, and in others it's heading back down again. This kind of heat-driven circulation is called convection.

According to physicists, a moving fluid can develop a magnetic field provided three conditions hold. First, the fluid must be able to conduct electricity – which iron can do fine. Secondly, there has to be at least a tiny magnetic field present to begin with – and there are good reasons to suppose that the Earth had a bit of personal magnetism, even early on. Thirdly, something has to *twist* the fluid, distorting that initial magnetic field – and for the Earth this twisting happens by way of Coriolis forces, which are like centrifugal forces but a bit more subtle, caused by the Earth's rotation on its axis. Roughly speaking, the twisting tangles the original, weak magnetic field like spaghetti being twirled on to a fork; then the magnetism bubbles upwards, trapped in the rising parts of the iron core. As a result of these motions, the magnetic field becomes a lot stronger.

So, yes, the Earth does behave a *bit* as though it had a huge bar magnet buried inside it, but there's rather more going on than that. Just to paint the picture in a little more detail, there are at least seven other factors that contribute to the Earth's magnetic field. Some of the materials of the Earth's crust *can* form permanent magnets. Like a compass needle pointing north, these materials align themselves with the stronger field from the geomagnetic dynamo and reinforce it. In the upper regions of the atmosphere is a layer of ionized gas – gas bearing an electrical charge. Until satellites were invented, this 'ionosphere' was crucial for radio communications, because radio waves bounced back down off the charged gas instead of beaming off into space. The ionosphere is moving, and moving electricity creates a magnetic field. About 15,000 miles (24,000 km) out lies the ring current, a low-density region of ionized particles forming a huge torus. This slightly reduces the strength of the magnetic field. The next two factors, the magnetopause and the magnetotail, are created by the interaction of the Earth's magnetic field with the solar wind – a continual stream of particles outward bound from our hyperactive sun. The magnetopause is the 'bow wave' of the Earth's magnetic field as it heads into the solar wind; the magnetotail is the 'wake' on the far side of the Earth, where the Earth's own field streams outwards getting ever more broken up by the solar wind. The solar wind also causes drag along the direction of the Earth's orbit, creating a further kind of motion of magnetic field lines known as field-aligned currents. Finally, there are the convective electrojets. The 'northern lights', or aurora borealis, are dramatic, eerie sheets of pale light that waft and shimmer in the northern polar skies: there is a similar display, the aurora australis, near the south pole. The auroras are generated by two sheets of electrical current that flow from magnetopause to magnetotail; these in turn create magnetic fields, the westward and eastward electrojets.

Yes, *like* a bar magnet – in the sense that an ocean is like a bowl of water.

Magnetic materials found in ancient rocks show that every so often – about once every half a million years, but with no sign of regularity – the Earth's magnetic field flips polarity, reversing magnetic north and south. We're not sure exactly why, but mathematical models suggest that the magnetic field can exist in these two orientations, with neither of them being totally stable. So whichever one it's in, it eventually loses stability and flips to the other one. The flips are rapid, taking about 5,000 years; the periods between flips are about a hundred times as long.

Most of the other planets have magnetic fields, and these can be even more complicated and difficult to explain than that of the Earth. We've still got a lot to learn about planetary magnetism.

One of the most dramatic features of our planet was discovered in 1912 but wasn't accepted by science until the 1960s, and some of the most compelling evidence was left by those flips in the Earth's magnetism. This is the notion that the continents are not fixed in place, but wander slowly over the surface of the planet. According to Alfred Wegener, the German who first publicized the idea, all of today's separate continents were originally part of a single super-continent, which he named Pangea ('All-Earth'). Pangea existed about 300 million years ago.

Wegener surely wasn't the first person to speculate along such lines, because he got the idea – in part, at least – from the curious similarity between the shapes of the coasts of Africa and South America. On a map the resemblance is striking. That wasn't Wegener's only source of inspiration, however. He wasn't a geolo-gist; he was a meteorologist, specializing in ancient climates. Why, he wondered, do we nowadays find rocks in regions with cold cli-mates that were clearly laid down in regions with warm climates? And why, for that matter, do we nowadays find rocks in regions with warm climates that were clearly laid down in regions with cold cli-mates? For example, remains of ancient glaciers 420 million years old can still be seen in the Sahara Desert, and fossil ferns are found in Antarctica. Pretty much everyone else thought that the climate must have changed: Wegener became convinced that the climate had stayed much the same, give or take the odd ice age, and the con-tinents had shifted. Perhaps they'd been driven apart by convection in the mantle – he wasn't sure.

This was considered a crazy idea: it wasn't suggested by a geol-ogist, *and* it ignored all sorts of inconvenient evidence, *and* the alleged fit between South America and Africa wasn't all that good anyway, *and* – to top it all – there was no conceivable mechanism for carting continents around. Certainly not convection, which was too weak. Great A'Tuin may lug a planet around on its back, but that's fantasy: in the real world, there seemed to be no conceivable way for it to happen.

We use the word 'conceivable' because a number of very bright and very reputable scientists were busily making one of the sub-ject's worst, and commonest, errors. They were confusing 'I can't see a way for this to happen' with 'There *is* no way for this to happen.' One of them, it pains one of us to admit, was a mathematician, and a brilliant one, but when his calculations told him that the Earth's mantle couldn't support forces strong enough to move continents, it didn't occur to him that the theories on which those calculations were based might be wrong. His name was Sir Harold Jeffreys, and he really should have been more imaginative, because it wasn't just the *shapes* of the land on either side of the Atlantic that fitted. The geology fitted too, and so did the fossil record. There is, for example, a fossil beast called *Mesosaurus*. It lived 270 million years ago, and is found only in South America and Africa. It couldn't have swum the Atlantic, but it could have evolved on Pangea and spread to both continents before they drifted apart.

In the 1960s, however, Wegener's ideas became orthodox and the theory of 'continental drift' became established. At a meeting of leading geologists, a Ponder Stibbons-like young man named Edward Bullard and two colleagues enlisted the aid of a new piece of kit called a computer. They instructed the machine to find the *best* fit between Africa and South America, *and* North America, *and* Europe, allowing for a bit of breakage but not too much. Instead of using today's coastline, which was never a very sensible idea but made it possible to claim that the fit wasn't actually that good, they used the contour corresponding to a depth of 3200 feet (1000 m) underwater, whose shape is less likely to have been changed by erosion. The fit *was* good, and the geology across the join matched amazingly well. And even though the people at the conference came out just as divided in their opinions as they'd been when they went in, somehow continental drift had become the consensus.

Today we have much more evidence, and a fair idea of the mechanism. Down the middle of the Atlantic Ocean, and elsewhere in other oceans, there runs a ridge – roughly north–south and about midway between South America and Africa. Volcanic material is welling up along that ridge, and spreading sideways. It's been spreading for 200 million years, and it's still doing it today: we can even send deep-sea submarines down there to watch. It's not spreading at speeds humans can see – America moves about three-quarters of an inch (2 cm) further away from Africa every year, about the same rate that your fingernails grow – but today's instruments can easily measure such a change.

The most striking evidence for continental drift is magnetic: the rocks on either side bear a curious pattern of magnetic stripes, reversing polarity from north to south and back again, and that pattern is *symmetric* on either side of the ridge – making it clear that the stripes were frozen in place as the rocks cooled in the Earth's magnetic field. Whenever the Earth's dynamo flipped polarity, as it does from time to time, the rock immediately adjacent to the ridgeline, on either side, got the same new polarity. As the rocks then spread apart, they took the same patterns of stripes with them.

The surface of the Earth is not a solid sphere. Instead, the continents and the ocean-beds float on top of large, essentially solid plates, and those plates can be driven apart by upwelling magma. (Oh, but mostly by convection in the mantle. Jeffreys didn't know what we now know about how the mantle moves.) There are about a dozen plates, ranging from 600 miles (1000 km) across to 6000 miles (10,000 km), and they twist and turn. Where plate boundaries rub against each other, sticking and slipping and sticking and slipping, you get a lot of earthquakes and volcanoes. Especially along the 'Pacific rim', the edge of the Pacific Ocean up along the west coast of Chile, central America, the USA, along down past Japan, and back round New Zealand, which is all the edge of a single gigantic plate. Where plate boundaries collide you get mountain ranges: one plate burrows under the other, lifting it up and crushing and folding its edges. India was once not part of the main Asian continent at all, but came crashing into it, creating the world's highest mountain range, the Himalayas. India hasn't fully stopped even now, and the Himalayas are still being pushed up by the force of the impact.

SEVENTEEN
SUIT OF SPELLS

 A FIGURE WAS FROGMARCHED through the early-morning corridors, surrounded by the senior wizards. It wore a long white nightshirt, and a nightcap with the word 'Wizzard' embroidered, inexpertly, on it. It was Unseen University's least qualified but most well-travelled member, usually away from something. And it was in trouble.

'This won't hurt a bit,' said the Senior Wrangler.

'It's right up your street,' said the Lecturer in Recent Runes.

'It's on a log and in your face,' explained the Dean.

'That isn't what HEX said, is it?' said the Senior Wrangler, as the sleepy figure was hustled around a corner.

'Very similar, but what HEX said made less sense,' said the Dean.

They hurried across the lawn and barged through the doors in the High Energy Magic Building.

Mustrum Ridcully finished filling his pipe, and struck a match on the dome of the Project. Then he turned, and smiled.

'Ah, Rincewind,' he said. 'Good of you to come.'

'I was dragged, sir.'

'Well done. And I have good news. I intend to appoint you Egregious Professor of Cruel and Unusual Geography. The post is vacant.'

Rincewind looked past him. On the far side of the room some of the junior wizards were working in a haze of magic that made it hard to see exactly what it was they were working on, but it looked almost like . . . some sort of skeleton.

'Oh,' he said. 'Er . . . but I'm very happy as assistant librarian. I'm getting really good at peeling the bananas.'

'But the new post offers you room, board and all your laundry done,' said the Archchancellor.

'But I get that already, sir.'

Ridcully drew leisurely on his pipe and blew out a cloud of blue smoke.

'Up until now,' he said.

'Oh. I see. And you're about to send me somewhere really dangerous, yes?'

Ridcully beamed. 'How *did* you guess?'

'It wasn't a guess.'

Fortunately the Dean had been forewarned and had grabbed the back of Rincewind's nightshirt, and so he was ready. The wizard's bedroom slippers skidded uselessly on the tiles as he tried to make for the door.

'It's best to let him run for a little while,' said the Senior Wrangler. 'It's a nervous reaction.'

'And the best thing is,' said Ridcully, to Rincewind's back, 'that although we are sending you to a place of immense danger where no living thing could possibly survive, you will not, in so many words, actually *be* there. Won't that be nice?'

Rincewind hesitated.

'How many words?'

'It'll be like being in a ... story,' said the Archchancellor. 'Or ... or a dream, as far as I can understand it. Mister Stibbons! Come and explain!'

'Oh, hello, Rincewind,' said Ponder, stepping out of the mist and wiping his hands on a rag. 'Twelve spells HEX has amalgamated for this! It's an amazing piece of thaumaturgical engineering! Do come and see!'

There are creatures which have evolved to live in coral reefs and simply could not survive in the rough, tooth-filled wastes of the open sea. They continue to exist by lurking among the dangerous tentacles of the sea anemone or around the lips of the giant clam and other perilous crevices shunned by all sensible fish.

A university is very much like a coral reef. It provides calm waters and food particles for delicate yet marvellously constructed organisms that could not possibly survive in the pounding surf of reality, where people ask questions like 'Is what you do of any use?'

and other nonsense.

In fact Rincewind in his association with UU had survived dangers that would have stripped a hero to the bone, but he nevertheless believed, despite all the evidence to the contrary, that he was *safe* in the university. He would do anything to stay on the roll.

At the moment this involved looking at some sort of skeletal armour made out of smoke while Ponder Stibbons gabbled incomprehensible words in his ear. As far as he could understand it, the thing put all your senses somewhere else when you stayed here. So far, that sounded quite acceptable, since it had always seemed to Rincewind that if you had to go a long way away it'd be nice to stay at home while you did it, but people seemed a little unclear about where *pain* fitted in.

'We'll send you – that is, your senses – somewhere,' said Ridcully.

'Where?' said Rincewind.

'Somewhere amazing,' said Ponder. 'We just want you to tell us what you see. And then we'll bring you back.'

'At what point will things go wrong?' said Rincewind.

'Nothing can possibly go wrong.'

'Oh,' Rincewind sighed. There was no point in arguing with a statement like that. 'Could I have some breakfast first?'

'Of course, dear fellow,' said Ridcully, patting him on the back. 'Have a hearty meal!'

'Yes, I thought that'd probably be the case,' said Rincewind gloomily.

When he'd been taken away, under escort by the Dean and a couple of college porters, the wizards clustered around the project.

'We've found a suitably large "sun", sir,' said Ponder, taking care to annunciate the inverted commas. 'We're moving the world now.'

'A very suspicious idea, this,' said the Archchancellor 'Suns go *around*. We see it happen every day. It's not some kind of optical illusion. This is a bit of a house of cards we're building here.'

'It's the only one available, sir.'

'I mean, things fall down because they're heavy, you see? The thing that causes them to fall down because they're heavy is, in fact, the fact that they're heavy. 'Heavy' means inclined to fall down. And, while you can call me Mr Silly –'

'Oh, I wouldn't do that, sir,' Ponder said, glad that Ridcully couldn't see his face.

'– I somehow feel that a crust of rock floating around on a ball of red-hot iron should not be thought of as "solid ground".'

'I think, sir, that this universe has a whole parcel of rules that take the place of narrativium,' said Ponder. 'It's ... sort of ... copying us, as you so perspicaciously pointed out the other day. It's making the only kind of suns that can work in it, and the only worlds that can exist if you don't have chelonium.'

'Even so ... going *around* a sun ... that's the sort of thing the Omnian priests used to teach, you know. Mankind is so insignificant that we just float around on some speck, and all that superstitious stuff. You know they used to persecute people for saying the turtle existed? And any fool can *see* it exists.'

'Yes, sir. It certainly does.'

There were problems, of course

'Are you sure it's the right sort of sun?' said Ridcully.

'You told HEX to find one that was "nice and yellow, nice and dull, and not likely to go off bang", sir,' said Ponder. 'It seems to be a pretty average one for this universe.'

'Even so ... tens of millions of miles ... that's a long way away for our world.'

'Yes, sir. But we tried some experimental worlds close to and they fell in, and we tried one a bit further out and that's baked like a biscuit, and there's one ... well, it's a bit of an armpit, really. The students have got quite good at making different sorts. Er ... we're calling them planets.'

'A planet, Stibbons, is a lump of rock a mere few hundred yards across which gives the night sky a little, oh, I don't know, what's the word, a little *je ne sais quoi* –'

'These will work, sir, and we've such a lot of them. As I said, sir, I've come to agree with your theory that, within the Project, matter is trying to do all by itself what in the real world is done by purpose, probably conveyed via narrativium.'

'Was that my theory?' said Ridcully.

'Oh, yes, sir,' said Ponder, who was learning the particular survival skills of the academic reef.

'It sounds rather a parody to me, but I dare say we will understand the joke in time. Ah, here comes our explorer. 'Morning, Professor,' said Ridcully. 'Are you ready?'

'No,' said Rincewind.

'It's very simple,' said Ponder, leading the reluctant traveller across the floor. 'You can think of this assemblage of spells as a suit of very, very good armour. Things will flicker , and then you'll be … somewhere else. Except you'll really be *here*, you see? But everything you see will be somewhere else. Absolutely nothing will hurt you because HEX will buffer all extreme sensations and you'll simply receive a gentle analogue of them. If it's freezing you'll feel rather chilly, if it's boiling you'll feel a little hot. If a mountain falls on you it'll be a bit of a knock. Time where you're going is moving very fast but HEX can slow it down while you are there. HEX says that he can probably exert small amounts of force within the Project, so you will be able to lift and push things, although it will feel as though you're wearing very large gloves. But this should not be required because all we want you to do to start with … Professor … is tell us what you see.'

Rincewind looked at the suit. It was, being largely make up of spells under HEX's control, shimmery and insubstantial. Light reflected off it in odd ways. The helmet was far too large and completely covered the face.

'I have three … no, four … no, *five* questions,' he said.

'Yes?'

'Can I resign?'

'No.'

'Do I have to understand anything you just told me?'

'No.'

'Are there any monsters where I'm being sent?'

'No.'

'Are you sure?'

'Yes.'

'Are you totally positive about that?'

'Yes.'

'I've just thought of another question,' said Rincewind.

'Fire away.'

'Are you *really* sure?'

'Yes!' snapped Ponder. 'And even if there were any monsters, it wouldn't matter.'

'It'd matter to me.'

'No it wouldn't! I have explained! If some huge toothed beast came galloping towards you, it'd have no effect on you at all.'

'Another question?'

'Yes?'

'Is there a toilet in this suit?'

'No.'

'Because there will be if a huge toothed beast comes galloping towards me.'

'In that case, you just say the word and you can come back and use the privy down the hall,' said Ponder. 'Now, stop worrying, please. These gentlemen will help you, er, insert yourself into the thing, and we'll begin ...'

The Archchancellor wandered up as the reluctant professor was enveloped in the glittering, not-quite-there stuff.

'A thought occurs, Ponder,' he said.

'Yes, sir?'

'I suppose there's no chance that there *is* life anywhere in the Project?'

Ponder looked at him in frank astonishment.

'Absolutely *not*, sir! It can't happen. Simple matter is obeying a few rather odd rules. That's probably enough to get things ... spinning and exploding and so on, but there's no possibility that they could cause anything so complex as –'

'The Bursar, for example?'

'Not even the Bursar, sir.'

'He's not very complicated, though. If only we could find a parrot that was good at sums, we could pension the old chap off.'

'No, sir. There's nothing like the Bursar. Not even an ant or a blade of grass. You might as well try to tune a piano by throwing rocks at it. Life does not turn up out of nowhere, sir. Life is a lot more than just rocks moving in circles. The one thing we're not going to run into is monsters.'

Two minutes later Rincewind blinked and found, when he opened his eyes, that they were somewhere else. There was a rather grainy redness in front of them, and he felt rather warm.

'I don't think it's working,' he said.

'You should be seeing a landscape,' said Ponder, in his ear.

'It's all just red.'

There was the sound of distant whispering. Then the voice said, 'Sorry. The aim wasn't very good. Wait a moment and we'll soon have you out of that volcanic vent.'

In the HEM Ponder took the ear trumpet away from his ear. The other wizards heard it sizzling, as if a very angry insect was trapped therein.

'Curious language,' he said, in mild surprise. 'well, let's raise him somewhat and let time move on a little ...'

He put the trumpet to his ear and listened.

'He says it's pissing down,' he announced.

EIGHTEEN
AIR AND WATER

 IT'S CERTAINLY A SURPRISE that the rigid rules of physics permit anything as flexible as life, and the wizards can hardly be blamed for not anticipating the possibility that living creatures might come into being on the barren rocks of Roundworld. But Down Here is not as different from Up There as it seems. Before we can talk about life, though, we need to deal with a few more features of our home planet: atmosphere and oceans. Without them, life as we know it could not have arisen; without life as we know it, our oceans and atmosphere would be distinctly different.

The story of the Earth's atmosphere is inextricably intertwined with that of its oceans. Indeed, the oceans can reasonably be viewed as just a rather damp, dense layer of the atmosphere. The oceans and the atmosphere evolved together, exerting strong influences on each other, and even today such an 'obviously' atmospheric phenomenon as weather turns out to be closely related to what happens in the oceans. One of the main recent breakthroughs in weather prediction has been to incorporate the oceans' ability to absorb, transport, and give off heat and moisture. To some extent, the same point can be made about the solid regions of the Earth, which also co-evolved with the air and the seas, and also interact with them. But the link between oceans and atmosphere is stronger.

The Earth and its atmosphere condensed together out of the primal gascloud that gave rise to the Sun and to the solar system. As a rough rule of thumb, the denser materials sank to the bottom of the condensing clump of matter that we now inhabit, and the lighter ones floated to the top. Of course there was, and still is, a lot more going on than that, so the Earth is not just a series of concentric shells of lighter and lighter matter, but the general distribution of solids, liquids, and gases makes sense if you think about it that way. And so, as the molten rocks of Earth began to cool and solid-ify, the nascent planet found itself already enveloped in a primordial atmosphere.

It was almost certainly very different from the atmosphere today, which is a mixture of gases, the main ones being the elements nitro-gen, oxygen and the inert gas argon, and the compounds carbon dioxide and water (in the form of vapour). The primordial atmos-phere also differed considerably from the gas cloud out of which it condensed – it wasn't just a representative sample of what was around. There are several reasons for this. One is that a solid planet and a gas cloud retain different gases. Another is that a solid planet can generate gases, by chemical or even nuclear reactions, or by other physical processes, which can escape from its interior into its atmosphere.

The early cloud was rich in hydrogen and helium, the lightest of elements. The speed with which a molecule moves becomes slower as the molecule gets heavier – a molecule with one hundred times the mass moves at about one-tenth the speed. Anything that moves faster than the Earth's escape velocity, about 7 miles per second (11 km/sec), can overcome the planet's gravity and disappear into space. Molecules in the atmosphere whose molecular weight – what you get by adding up the atomic weights of the component atoms – is less than about 10 should therefore disappear into the void. Hydrogen has molecular weight 2, helium 4, so neither of these otherwise abundant gases should be expected to hang around. The most abundant molecules in the primal gas cloud, with molecular weight greater than 10, are methane, ammonia, water, and neon. This is similar to what we find today on the gas giants Jupiter, Saturn, Uranus, and Neptune – except that they are more massive, so have a greater escape velocity, and can retain lighter gases such as hydrogen and helium as well. We can't be certain that the Earth of 4 billion years ago possessed a methane-ammonia atmosphere, because we don't know exactly how the primal gas cloud condensed, but it is clear that if the ancient Earth ever possessed such an atmos-phere, it lost nearly all of it. Today there is little methane or ammonia, and what there is has a biological origin.

Shortly after the Earth was formed, the atmosphere contained very little oxygen. Around 2 billion years ago, the proportion of oxygen in the atmosphere increased to about 5%. The most likely cause of this change – though perhaps not the only one – was the evolution of photosynthesis. At some stage, probably soon after 4 billion years ago, bacteria in the oceans evolved the trick of using the energy of sunlight to turn water and carbon dioxide into sugar and oxygen. The oxygen that they produced did not show up in the atmosphere, in any appreciable amount, until two billion years ago. A lot of gases and minerals had to be oxidised first. Plants use the same trick today, and they use the same molecules as one of the early bacteria did: chlorophyll. Animals proceed in pretty much the opposite direction: they power themselves by using oxygen to burn food, producing carbon dioxide instead of using it up. Those early photosynthesizing bacteria used the sugar for energy, and multiplied rapidly, but to them the oxygen was just a form of toxic waste, which bubbled up into the atmosphere. The oxygen level then stayed roughly constant until about 600 million years ago, when it underwent a rapid increase to the current level of 21%.

The amount of oxygen in today's atmosphere is far greater than could ever be sustained without the influence of living creatures, which not only produce oxygen in huge quantities but use it up again, in particular locking it up in carbon dioxide. It is startling how far 'out of balance' the atmosphere is, compared to what would happen if life were suddenly removed and only inorganic chemical processes could act. The amount of oxygen in the atmosphere is dynamic – it can change on a timescale that by geological standards is extremely rapid, a matter of centuries rather than millions of years. For example, if some disaster occurred which killed off all the plants but left all the animals, then the proportion of oxygen would *halve* in about 500 years, to the level on mountain peaks in the Andes today. The same goes for the scenario of 'nuclear winter' introduced by Carl Sagan, in which clouds of dust thrown into the atmosphere by a nuclear war stop most of the sunlight from reaching the ground. In this case, plants may still eke out some kind of existence, but they don't photosynthesize: they do use oxygen, though, and so do the microorganisms that break down dead plants.

The same screening effect could also occur if there were unusual numbers of active volcanoes, or if a big meteorite or comet hit the Earth. When comet Shoemaker-Levy 9 hit Jupiter in 1994, the impact was equivalent to half a million hydrogen bombs.

The 'budget' of income and expenditure for oxygen, and the associated but distinct budget for carbon, is still not understood. This is an enormously important question because it is vital background to the debate about global warming. Human activities, such as electrical power plants, industry, use of cars, or simply going about one's usual business and breathing while one does so, generate carbon dioxide. Carbon dioxide is a 'greenhouse gas' which traps incoming sunlight like the glass of a greenhouse. So if we produce too much carbon dioxide, the planet should warm up. This would have undesirable consequences, ranging from floods in low-lying regions such as Bangladesh to big changes in the geographical ranges of insects, which could inflict serious damage on crops. The question is: do these human activities actually increase the Earth's carbon dioxide, or does the planet compensate in some way? The answer makes the difference between imposing major restrictions on how people in developed (and developing) countries live their lives, and letting them continue along their current paths. The current consensus is that there are clear, though subtle, signs that human activities *do* increase the carbon dioxide levels, which is why major international treaties have been signed to reduce carbon dioxide output. (Actually taking that action, rather than just promising to do so, may prove to be a different matter altogether.)

The difficulties involved in being sure are many. We don't have good records of past levels of carbon dioxide, so we lack a suitable 'benchmark' against which to assess today's levels – although we're beginning to get a clearer picture thanks to ice cores drilled up from the Arctic and Antarctic, which contain trapped samples of ancient atmospheres. If 'global warming' *is* under way, it need not show up as an increase in temperature anyway (so the name is a bit silly). What it shows up as is climatic disturbance. So even though the eight warmest summers in Britain in the 20th century all occurred in the nineties, we can't simply conclude that 'it's getting warmer', and hence that global warming is a fact. The global climate varies wildly anyway – what would it be doing if we weren't here?

A project known as Biosphere 2 attempted to sort out the basic science of oxygen/carbon transactions in the global ecosystem by setting up a 'closed' ecology – a system with no inputs, beyond sunlight, and no outputs whatsoever. In form it was like a gigantic futuristic garden centre, with plants, insects, birds, mammals, and people living inside it. The idea was to keep the ecology working by choosing a design in which everything was recycled.

The project quickly ran into trouble: in order to keep it running, it was necessary to keep adding oxygen. The investigators therefore assumed that somehow oxygen was being lost. This turned out to be true, in a way, but for nowhere near as literal a reason. Even though the whole idea was to monitor chemical and other changes in a closed system, the investigators hadn't weighed how much carbon they'd introduced at the start. There were good reasons for the omission – mostly, it's extremely difficult, since you have to estimate carbon content from the wet weight of live plants. Not knowing how much carbon was really there to begin with, they couldn't keep track of what was happening to carbon monoxide and carbon dioxide. However, 'missing' oxygen ought to show up as increased carbon dioxide, and they could monitor the carbon dioxide level and see that it wasn't going up.

Eventually it turned out that the 'missing' oxygen wasn't escaping from the building: it was being turned into carbon dioxide. So why didn't they see increased carbon dioxide levels? Because, unknown to anybody, carbon dioxide was being absorbed by the building's concrete as it 'cured'. Every architect knows that this process goes on for ten years or so after concrete has set, but this knowledge is irrelevant to architecture. The experimental ecologists knew nothing about it at all, because esoteric properties of poured concrete don't normally feature in ecology courses, but to them the knowledge was vital.

Behind the unwarranted assumptions that were made about Biosphere 2 was a plausible but irrational belief that because carbon dioxide *uses up* oxygen when it is formed, then carbon dioxide is *opposite* to oxygen. That is, oxygen counts as a credit in the oxygen budget, but carbon dioxide counts as a debit. So when carbon dioxide disappears from the books, it is interpreted as a debt cancelled, that is, a credit. Actually, however, carbon dioxide contains a positive quantity of oxygen, so when you lose carbon dioxide you lose oxygen too. But since what you're looking for is an increase in carbon dioxide, you won't notice if some of it is being lost.

The fallacy of this kind of reasoning has far wider importance than the fate of Biosphere 2. An important example within the general frame of the carbon/oxygen budget is the role of rainforests. In Brazil, the rainforests of the Amazon are being destroyed at an alarming rate by bulldozing and burning. There are many excellent reasons to prevent this continuing – loss of habitat for organisms, production of carbon dioxide from burning trees, destruction of the culture of native Indian tribes, and so on. What is *not* a good reason, though, is the phrase that is almost inevitably trotted out, to the effect that the rainforests are the 'lungs of the planet'. The image here is that the 'civilized' regions – that is, the industrialized ones – are net producers of carbon dioxide. The pristine rainforest, in contrast, produces a gentle but enormous oxygen breeze, while absorbing the excess carbon dioxide produced by all those nasty people with cars. It *must* do, surely? A forest is full of plants, and plants produce oxygen.

No, they don't. The net oxygen production of a rainforest is, on average, zero. Trees produce carbon dioxide at night, when they are not photosynthesizing. They lock up oxygen and carbon into sugars, yes – but when they die, they rot, and release carbon dioxide. Forests can indirectly remove carbon dioxide by removing carbon and locking it up as coal or peat, and by releasing oxygen into the atmosphere. Ironically, that's where a lot of the human production of carbon dioxide comes from – we dig it up and burn it again, using up the same amount of oxygen.

If the theory that oil is the remains of plants from the carboniferous period is true, then our cars are burning up carbon that was once laid down by plants. Even if an alternative theory, growing in popularity, is true, and oil was produced by bacteria, then the problem remains the same. Either way, if you burn a rainforest you add a one-off surplus of carbon dioxide to the atmosphere, but you do *not* also reduce the Earth's capacity to generate new oxygen. If you want to reduce atmospheric carbon dioxide *permanently*, and not just cut short-term emissions, the best bet is to build up a big library at home, locking carbon into paper, or put plenty of asphalt on roads. These don't sound like 'green' activities, but they are. You can cycle on the roads if it makes you feel better.

Another important atmospheric component is nitrogen. It is a lot easier to keep track of the nitrogen budget. Organisms – plants especially, as every gardener knows – need nitrogen for growth, but they can't just absorb it from the air. It has to be 'fixed' – that is, combined into compounds that organisms can use. Some of the fixed nitrogen is produced as nitric acid, which rains down after thunderstorms, but most nitrogen fixation is biological. Many simple lifeforms 'fix' nitrogen, using it as a component of their own amino-acids. These amino-acids can then be used in everybody else's proteins.

The Earth's oceans contain a huge quantity of water – about a third of a billion cubic miles (1.3 billion cubic km). How much water there was in the earliest stages of the Earth's evolution, and how it was distributed over the surface of the globe, we have little idea, but the existence of fossils from about 3.3 billion years ago shows that there must have been water around at that time, probably quite a lot. As we've already explained, the Earth – along with the rest of the solar system, Sun included – condensed from a vast cloud of gas and dust, whose main constituent was hydrogen. Hydrogen combines readily with oxygen to form water, but it also combines with carbon to form methane and with nitrogen to form ammonia.

The primitive Earth's atmosphere contained a lot of hydrogen and a fair quantity of water vapour, but initially the planet was too hot for liquid water to exist. As the planet slowly cooled, its surface passed a critical temperature, the boiling point of water. That temperature was probably not exactly the same as the one at which

water boils now; in fact even today it's not one inflexible temperature, because the boiling point of water depends on pressure and
other circumstances. Nor was it just a simple matter of the atmosphere's getting colder: its composition also changed because the
Earth was spouting out gases from its interior through volcanic
activity.

A crucial factor was the influence of sunlight, which split some
of the atmospheric water vapour into oxygen and hydrogen. The
hydrogen escaped from the Earth's relatively weak gravitational
field, so the proportion of oxygen got bigger while that of water
vapour got smaller. The effect of this was to *increase* the temperature
at which the water vapour could condense. So as the temperature of
the atmosphere slowly fell, the temperature at which water vapour
would condense rose to meet it. Eventually the atmosphere going
down passed the boiling point of water going *up*, and water vapour
began to condense into liquid water ... and to fall as rain.

It must have absolutely bucketed down.

When the rain hit the hot rocks beneath, it promptly evaporated
back into vapour, but as it did so it cooled the rocks. Heat and temperature are not the same. Heat is equivalent to energy: when you
heat something, you input extra energy. Temperature is one of the
ways in which that energy can be expressed: it is the vibration of
molecules. The faster those vibrations are, the higher the temperature. Ordinarily, the temperature of a substance goes up if you heat
it: all the extra heat is expressed as more vibration of the molecules.
However, at transitions from solid to liquid, or liquid to vapour or
gas, the extra heat goes into changing the state of the substance, not
into making its temperature higher. So you can throw in a lot of heat
and instead of the stuff getting hotter, it changes state – a so-called
phase transition. Conversely, when a substance cools through a
phase transition, it gives off a lot of heat. So the cooling water
vapour put more heat back into the upper atmosphere, from which
it could be radiated away into space and lost. When the hot rocks
turned the water back into vapour, the rocks got a lot cooler very
suddenly. In a geologically short space of time, the rocks had cooled
below the boiling point of water, and now the falling rain no longer

got turned back into vapour – at least, not much of it did.

It may well have rained for a million years. So it's not surprising that Rincewind noticed that it was a bit wet.

Thanks to gravity, water goes downhill, so all that rain accumulated in the lowest depressions in the Earth's irregular surface. Because the atmosphere had a lot of carbon dioxide in it, those early oceans contained a lot of dissolved carbon dioxide, making the water slightly acidic. There may have been hydrochloric and sulphuric acids too. The acid ate away at the surface rocks, causing minerals to dissolve in the oceans; the sea began to get salty.

At first the amount of oxygen in the atmosphere increased slowly, because the effect of incoming sunlight isn't particularly dramatic. But now life got in on the act, bubbling off oxygen as a by-product of photosynthesis. The oxygen combined with any remaining hydrogen in the atmosphere, whether on its own or combined inside methane, to produce *more* water. This also fell as rain, and increased the amount of ocean, leading to more bacteria, more oxygen – and so it continued until the available hydrogen pretty much ran out.

Originally it used to be thought that the oceans just kept dissolving the rocks of the continents, accumulating more and more minerals, getting saltier and saltier until the amount of salt reached its current value of about 3.5%. The evidence for this is the percentage of salt in the blood of fishes and mammals, which is about 1%. In effect, it was believed that fish and mammal blood were 'fossilized' ocean. Today we are still often told that we have ancient seas in our blood. This is probably wrong, but the argument is far from settled. It is true that our blood is salty, and so is the sea, but there are plenty of ways for biology to adjust salt content. That 1% may just be whatever level of salt makes best sense for the creature whose blood it is. Salt – more properly, the ions of sodium and chlorine into which it decomposes – have many biological uses: our nervous systems, for instance, wouldn't work without them. So while it is entirely believable that evolution took advantage of the *existence* of salt in the sea, it need not be stuck with the same proportion. On the other hand, there is good reason to think that cells first evolved as tiny free-floating organisms in the oceans, and those early cells weren't sophisticated enough to fight against a difference in salt concentration between their insides and their outsides, so they may well have settled on the same concentration because that was all they could initially manage – and having done so, they were rather stuck with it.

Can we decide by taking a more careful look at the oceans? Oceans have ways to lose salt as well as gaining it. Seas can dry out; the Dead Sea in Israel is a famous example. There are salt mines all over the place, relics of ancient dried-up seas. And just as living creatures – bacteria – took out carbon dioxide, turning it into oxygen and sugar, so they can take out other dissolved minerals too. Calcium, carbon and oxygen go into shells, for instance, which fall to the ocean floor when their owner dies. The clincher is ... time. The oceans are thought to have reached their current composition, and in particular their current degree of saltiness, about 2 to 1.5 billion years ago. The evidence is the chemical composition of sedimentary rocks – rocks formed from deposits of shells and other hard parts of organisms – which seems not to have changed much in the interim. (Though in 1998 Paul Knauth presented evidence that the early ocean may have been *more* salty than it is now, with somewhere between 1.5 to 2 times as much salt. His calculations indicate that salt could not have been deposited on the continents until about 2.5 billion years ago.) Simple calculations based on how much material dissolves in rivers and how fast rivers flow show that the entire salt content of the oceans can be supplied from dissolved continental rocks in twelve million years – the twinkling of a geological eye. If salt had just built up steadily, the oceans would now be far more salt than water. So the oceans are not simply sinks for dissolved minerals, one-way streets into which minerals flow and get trapped. They are mineral-processing machines. The geological evidence of the similarity of ancient and modern sedimentary rocks suggests that the inflow and the outflow pretty much balance each other.

So do we have ancient seas in our blood? In a way. The *proportions* of magnesium, calcium, potassium, and sodium are exactly the same as they were in the ancient seas from which our blood may have evolved – but cells seem to prefer a salt concentration of 1%, not 3%.

NINETEEN
THERE IS
A TIDE ...

 'HE'S RIGHT ABOUT THE RAIN,' said the Senior Wrangler, who was at the omniscope. 'You've got clouds again. And there's lots of volcanoes.'

'I'm moving him on further ... Oh. Now he says it's dark and cold and he's got a headache ...'

'Not very *graphic*, is it?' said the Dean.

'He says it's a splitting headache.'

HEX wrote something.

'Oh,' said Ponder. 'He's under water. I'm sorry about that, I'm afraid he's a little hard to position accurately. We're still not sure what size he should be. How's this?'

The trumpet rattled. 'He's still under water, but he says he can see the surface. I think that's as good as we're going to get. Just walk forward.'

As one wizard, they turned to watch the suit.

It hung in the air, a few inches above the floor. As they watched, the figure inside made hesitant walking motions.

It was not a nice day.

It was still raining, although it had slackened off recently, with sporadic outbreaks during the early part of the millennium and scattered showers during the last couple of decades. Now ten thousand rivers were finding their way to the sea. The light was grey and gave the beach a flat, monochrome, and certainly very damp look.

Whole religions have been inspired by the sight of a figure emerging, miraculously, from the sea. It would be hard to guess at what strange cult might be inspired by the thing now trudging out of the waves, although avoidance of strong drink and certainly of seafood would probably be high on its list of 'don'ts'.

Rincewind looked around.

166

There was no sand underfoot. The water sucked at an expanse of rough lava. There was no seaweed, no seabirds, no little crabs – nothing potentially dangerous at *all*.

'There's not a lot going on,' he said. 'It's all rather dull.'

'It'll be dawn in a moment,' said Ponder's voice in his ear. 'We'll be interested to see what you think of it.'

Strange way of putting it, Rincewind thought, as he watched the sun come up. It was hidden behind the clouds, but a greyish-yellow light picked its way across the landscape.

'It's all right,' he said. 'The sky's a dirty colour. Where is this? Llamedos? Hergen? Why aren't there *any* seashells? Is this high tide?'

All the wizards were trying to speak at once.

'I can't think of *everything*, sir!'

'But *everyone* knows about tides!'

'Perhaps some mechanism for raising and lowering the sea bed would be acceptable?'

'If it comes to that, what causes tides *here*?'

'Can we all please stop *shouting*?'

The babble died down.

'Good,' said Ridcully. 'Over to you, Mister Stibbons.'

Stibbons stared at the notes in front of him.

'I'm ... there's ... it's a puzzler, sir. On a round world the sea just sits there. There's no edge for it to pour off.'

'It's always been believed that the sea is in some way attracted to the moon,' the Senior Wrangler mused. 'You know ... the attraction of serene beauty and so on.'

Dead silence fell.

Finally, Ponder managed: 'No one said anything to me about a moon.'

'You've *got* to have a moon,' said Ridcully.

'It should be easy, shouldn't it?' said the Dean. '*Our* moon goes around the Disc.'

'But where can we put it?' said Ponder. 'It's got to be light and dark, we've got to move it for phases, and it's got to be almost as big as the sun and we *know* that if you try to make things sun-sized here they, well, become suns.'

'Our moon is closer than the sun,' said the Dean. 'That's why we get eclipses.'

'Only about ninety miles,' said Ponder. 'That's why it's burned black on one side.'

'Dear me, Mister Stibbons, I'm surprised at you,' said Ridcully. 'The damn great sun looks pretty big even though it's a long way away. Put the moon nearer.'

'We've still got the big lump that the Dean knocked out of the planet,' said the Senior Wrangler. 'I made the students park it around the Target.'

'Target?' said Ponder.

'It's the big fat planet with the coloured lines on,' said the Senior Wrangler. 'I made them bring the whole lot out to the new, er, sun because frankly they were a nuisance where they were. At least when they're spinning round you know where they're coming from.'

'Are the students still sneaking in here at night to play games?' said Ridcully.

'I've put a stop to that,' said the Dean. 'There's too many rocks and snowballs around this sun in any case. *Masses* of the things. Such a waste.'

'Well, can we get the lost lump here soon?'

'Hex can manipulate time from Rincewind's point of view,' said Ponder. 'For us, Project time is very fast ... we should get it here before the coffee arrives.'

'Can you hear me, Rincewind?'

'Yes. Any chance of some lunch?'

'We're getting you some sandwiches. Now, can you see the sun properly?'

'It's all very hazy, but yes.'

'Can you tell me what happens if I do ... this?'

Rincewind squinted into the grey sky. Shadows were racing across the landscape.

'You're not going to tell me you've just caused an eclipse of the sun, are you?'

Rincewind could hear faint cheering in the background.

'And you're quite certain it's an eclipse?' said Ponder.

'What else is it? A black disc is covering the sun and there's no birdsong.'

'Is it about the right size?'

'What kind of question is that?'

'All right, all right. Ah, here are your san – what? How? Excuse me ... *now* what? ...'

The senior wizards were puzzled again, and demonstrated this by prodding Ponder while he was trying to talk. The wizards were great ones for the prod as a means of getting attention.

'You can *see* there's only one moon,' said the Senior Wrangler, for the third time.

'All right ... how about this?' said Ponder. 'Let us suppose that in some way this world has got both water that *likes* moons and water that can't stand moons at any price. If it's got about the same amount of both, then that at least explains why there seem to be high tides on both sides at once. I think we can dispose of the Invisible Moon theory, interesting though it was, Dean.'

'I like that explanation,' said Ridcully. 'It is elegant, Mister Stibbons.'

'It's only a guess, sir.'

'Good enough for physics,' said Ridcully.

TWENTY

A GIANT LEAP
FOR MOONKIND

 HUMANITY HAS ALWAYS KNOWN the Moon is important. It often comes out at night, which is useful;
it changes, in a sky where change is rare; some of
us believe our ancestors live there. That last one
might not be capable of experimental verification,
but nevertheless humanity in general got it right.
The Moon reaches out ghostly tentacles, gravity and light; it may
even be our protector.

The wizards are right to worry that they've forgotten to give
Roundworld a Moon, though as usual they're worried for the
wrong reasons.

The Moon is a satellite of the Earth: we go round the Sun, but
the Moon goes round *us*. It's been up there for a long time, and in
its quiet way it's been exceedingly busy. The Moon affects people as
well as baby turtles. The main way it affects us is by causing tides.
It may affect us in other, less obvious ways, although many common
beliefs about the moon are, to say the least, scientifically controversial. The female menstrual cycle repeats roughly every four weeks,
much the same time that it takes the moon to go round the Earth –
one month, in fact, a word that comes from 'moon'. In popular
belief this numerical similarity is no coincidence, as for example in
'the wrong time of the month'. On the other hand, the Moon is the
epitome of regularity, as predictable as the date of Christmas day,
which cannot be said of the menstrual cycle.* Lovers, of course,
swoon and spoon beneath the Moon in June … It is also widely held

---

* Moreover, until the last few decades of human history, most women did not
cycle. Nearly all the time, they were either pregnant or lactating. And for the
great apes, the cycle is a week or so longer than for humans, and for gibbons
it's shorter. So it looks as though the relation with the Moon is coincidental.

that people go mad when there is a full Moon, or – a more extreme type of madness – those who are suitably afflicted turn into wolves for a night.

The werewolf legend plays a central role in *Men at Arms*. Most of the time lance-constable Angua of the Ankh-Morpork city watch is a well-built ash-blonde, but when the Moon is full she turns into a wolf who can smell colours and rip out people's jugular veins. But it *does* play havoc with her private life. 'It was always a problem, growing fangs and hair every full moon. Just when she thought she'd been lucky before, she'd found that few men are happy in a relationship where their partner grows hair and howls.' Fortunately Corporal Carrot is unperturbed by these occasional changes. He likes a girlfriend who enjoys long walks.

The Moon is unusual, and it is quite likely that without it, none of us would be here at all. Not because of the alleged effect on lovers, who find a way Moon or no, but because the Moon protects the Earth from some nasty influences that might have made it difficult for life to have arisen, or at least to have got beyond the most rudimentary forms. What makes the Moon unusual is not that it is a companion to a planet: all of the planets except Mercury and Venus have moons. It *is* remarkable because it is so big in comparison to its parent planet. Only Pluto has a satellite – Charon, discovered in 1978 by Jim Christy – that is comparable in relative size to our Moon. It's not stretching things much to say that we live on one half of a double planet.

We know the Moon is very different from the Earth in all sorts of ways. Its gravity is weaker, so it wouldn't be able to keep an atmosphere for very long, even if it had one, which it doesn't by any sensible use of the term. The Moon's surface is rock and rock dust, with no seas anywhere (water easily escapes too) – although in 1997 NASA probes discovered substantial quantities of water ice at the Moon's poles, hidden from the warmth of the Sun by the permanent shadows of crater walls. That's good news for future lunar colonies, which could act as bases for the exploration of the solar system. The Moon is a good place to start from, because your spaceship doesn't need much fuel to escape the Moon's pull; the

Earth is of course a bad place to start from, because down here gravity is so much stronger. How typical of humans to have evolved in the wrong place ...

How was the Moon formed? Did it condense out of the primal dustclouds along with the Earth? Did it form separately and get captured later? Are the craters extinct volcanoes, or are they marks made by lumps of rock smashing into the Moon? We know rather more about the Moon than we do about most other bodies in the solar system, because *we've been there*. In April 1969, Neil Armstrong stepped down on to the surface of the Moon, fluffed his lines, and made history. Between 1968 and 1972 the United States sent ten Apollo missions to the Moon and back. Of these, Apollos 8, 9, and 10 were never intended to land; Apollo–11 was that historic first landing; and Apollo–13 never made it down to the surface, suffering a disastrous explosion early in its flight and turning into an excellent movie.

The rest of Apollos 11-17 landed, and between them they brought back 800 lb (400 kg) of moon rock. Most of it is still stored in the Lunar Curatorial Facility in NASA's Johnson Space Center at Clear Lake, Houston; a lot of it has never been seriously looked at at all, but what *has* been analysed has taught us a lot about the origins and nature of the Moon.

The Moon is about a quarter of a million miles (400,000 km) from the Earth. It is less dense than the Earth, on average, but the Moon's density is very similar to that of the Earth's *mantle*, a curious fact that may not be coincidence. The same side of the Moon always faces the Earth, though it wobbles a bit. The dark markings on it are called *maria*, Latin for 'seas' – but they're not. They're flattish plains of rock which at one time was molten and flowed across the lunar surface like lava from a volcano. Nearly all of the craters are impact craters, where meteorites have smashed into the Moon. There are lots of them because there's a lot of rocks floating about in space, the Moon has no atmosphere to shield it by burning up the rocks through frictional heating, and the Moon has no weather to grind them back down again until they disappear. The Earth's atmosphere is a pretty good shield, but once geologists started looking they found remains of 160 impact craters down here, which is interesting given that a lot of them will have eroded away in the wind and the rain. But more of that when we get to dinosaurs.

Today the Moon always turns the same face to the Earth, which means that it rotates once round its axis every month, the same time that it takes to revolve around the Earth. (If it didn't rotate at all, it would always be pointing in the same direction – not the same direction relative to the Earth, but the same direction period. Imagine someone walking round you in a circle but always facing north, say. Then they don't always face *you*. In fact, you see all sides of them.) It wasn't always like this. Over hundreds of millions of years, the effect of tides has been to slow down the rotation rates of both Earth and Moon. Once the moon's rotation became synchronized with its revolutions round the Earth, the system stabilized. The moon also used to be quite a bit closer to the Earth, but over long periods of time it has moved further and further out.

Between 1600 and 1900 three theories of the formation of the Moon came into vogue and out again. One was that the Moon had formed at the same time as the Earth when the dustcloud condensed to form the solar system – Sun, planets, satellites, the whole ball of wax ... or rock, anyway. This theory, like early theories of the solar system's formation, falls foul of angular momentum. The Earth is spinning too fast, and the moon is revolving too fast, to be consistent with the Moon condensing from a dustcloud. (We misled you earlier when we said that the dustcloud theory explained the satellites too. Mostly it does, but not our enigmatic Moon. Lies-to-children, you see – *now* you're ready for the next layer of complication.)

Theory two was that the Moon is a piece of the Earth that broke away, maybe when the Earth was still completely molten and spinning rather fast. That theory bounced into the bin because nobody could find a plausible way for a spinning molten Earth to eject anything that would remotely resemble the Moon, even if you waited a bit for things to cool down.

According to theory three, the Moon formed elsewhere in the

solar system, and was wandering along when it happened to come within the Earth's gravitational clutches and couldn't get out again. This theory was very popular, even though gravitational capture is distinctly tricky to arrange. It's a bit like trying to throw a golfball into the hole so that it goes round and round just inside the rim. What usually happens is that it falls to the bottom (collides with the Earth) or does what every golfer has experienced to their utter horror, and goes in for a split second before climbing back out again (escapes without being captured).

The rock samples from Apollo missions added to the mystery of the Moon's origins. In some respects, Moon rock is astonishingly similar to Earth rock. If they were similar in most respects, this would be evidence for a common origin, and we'd have to take another look at the theory that they both condensed from the same dustcloud. But Moon rock doesn't resemble *all* Earth rock, only the mantle. The current theory, which dates from the early 1980s, is that the Moon was once *part* of the Earth's mantle. It wasn't ejected as a result of the Earth's spin: it was knocked into space about four billion years ago when a giant body, about the size of Mars, struck the early Earth a glancing blow. Computer calculations show that such an impact can, if conditions are right, strip a large chunk of mantle from the Earth, and sort of smear it out into space. This takes about 13 minutes (aren't computers good?). Then the ejected mantle, which is molten, begins to condense into a ring of rocks of various sizes. Some of it forms a big lump, the proto-Moon, and this quickly sweeps up most of the rest. What's left doesn't go away so easily, however, but over 100 million years nearly all of it crashes into either the Moon or the Earth, because of gravity.

The first simulations to support this theory of the Moon's formation suffered from some problems; in particular, they dated the impact very early in the Earth's formation, in order to get the Moon's angular momentum right. But if the collision had occurred that early, then the Moon would have accumulated a lot of iron from later impacts, just as the Earth did. However, there is little iron on (or in) the Moon. More recent work shows that a rather later impact could also give the Moon the correct angular momentum, and avoids this difficulty. However, it predicts that about 80% of the impacting body would have ended up on the Moon. In order for the Moon to resemble Earth's mantle as closely as it does, the impacting body would *also* have had to resemble the Earth's mantle.

This line of theorising may be losing the plot, though, because it was that very resemblance that needed explanation in the first place, and gave rise to the giant impact theory of the Moon's formation. Anything that explains why the impactor was like the mantle, such as 'it formed at the same distance from the Sun as the Earth did,' can probably explain why the Moon is like the mantle, without an impactor being needed. Maybe both the Moon and the Earth's mantle were splashed off something *else* by an impact.

Because Earth has weather – especially back then, oh boy, did it have weather *then* – the resulting impact craters all got eroded away; but because the Moon has no weather, the lunar impact craters didn't get eroded away, and a lot of them are still there now. The great charm of this theory is that it explains many different features of the Moon in one go – its similarity to the Earth's mantle, the fact that its surface seems to have undergone a sudden and extreme amount of heating about 4 billion years ago, its craters, its size, its spin – even those sea-like maria, released as the proto-Moon slowly cooled. The early solar system was a violent place.

In fact, the Dean's mis-designed sun might have done us some good after all …

The Moon affects life on Earth in at least two or three ways that we know of, probably dozens more that we haven't yet appreciated.

The most obvious effect of the Moon on the Earth is the tides – a fact that the wizards are stumbling towards. Like most of science, the story of the tides is not entirely straightforward, and only loosely connected to what common sense, left to its own devices, would lead us to expect. The common sense bit is that the Moon's gravity pulls at the Earth, and it pulls more strongly on the bit that is closest to the Moon. When that bit is land, nothing much happens, but when it's water – and more than half our planet's surface is ocean – it can pile up. This explanation is a lie-to-children, and it doesn't agree with what actually happens. It leads us to expect that at any given place on Earth, high tide occurs when the Moon is overhead, or at least at its highest point in the sky. That would lead to one high tide every day – or, allowing for a little complexity in the Earth–Moon system, one high tide every 24 hours 50 minutes.

Actually, high tides occur twice a day, 12 hours and 25 minutes apart. Exactly half the figure.

Not only that: the pull of the Moon's gravity at the surface of the Earth is only one ten millionth of the Earth's surface gravity; the pull of the Sun is about half that. Even when combined together, these two forces are not strong enough to lift masses of water through heights of up to 70 feet (21 m) – the biggest tidal movement on Earth, occurring in the Bay of Fundy between Nova Scotia and New Brunswick.

An acceptable explanation of the tides eluded humanity until Isaac Newton worked out the law of gravity and did the necessary calculations. His ideas have since been refined and improved, but he had the basics.

For simplicity, ignore everything except the Earth and the Moon, and assume that the Earth is completely made of water. The watery Earth spins on its axis, so it is subjected to centrifugal force and bulges slightly at the equator. Two other forces act on it: the Earth's gravity and the Moon's. The shape that the water takes up in response to these forces depends on the fact that water is a fluid. In normal circumstances, the surface of a standing body of water is horizontal, because if it wasn't, then the fluid on the higher bits would slosh sideways into the lower bits. The same kind of thing happens when there are extra forces acting: the surface of the water settles at right angles to the net direction of the combined forces.

When you work out the details for the three forces we've just mentioned, you find that the water forms an ellipsoid, a shape that is close to a sphere but very slightly elongated. The direction of elongation points towards the Moon. However, the centre of the ellipsoid coincides with the centre of the Earth, so the water 'piles up' on the side furthest from the Moon as well as on the side nearest it. This change of shape is only partly caused by the Moon's gravity 'lifting' the water closest to it. Most of the motion, in fact, is sideways rather than upwards. The sideways forces push more water into some regions of the oceans, and take it away from others. The total effect is *tiny* – the surface of the sea rises and falls through a distance of 18 inches (half a metre).

The coast, where land meets sea, is what creates the big tidal movements. Most of the water is moving sideways (not up) and its motion is affected by the shape of the coastline. In some places the water flows into a narrowing funnel, and then it piles up much more than it does elsewhere. This is what happens in the Bay of Fundy. This effect is made even bigger because coastal waters are shallow, so the energy of the moving water gets concentrated into a thinner layer, creating bigger and faster movements.

Finally, let's put the sun back. This has the same kind of effect as the Moon, but smaller. When Sun and Moon are aligned – either both on the same side of the Earth, in which case we see a new moon, or both on opposite sides (full moon) – their gravitational pulls reinforce each other, leading to so-called 'spring tides' in which high tide is higher than normal and low tide is lower. These have nothing to do with the *season* Spring. When the Sun and Moon are at right angles as seen from Earth, at half moon, the Sun's pull cancels out part of the Moon's, leading to 'neap tides' with less movement than normal (these presumably have nothing to do with the *season* Neap ...).

By putting all these effects together, and keeping good records of past tides, it is possible to predict the times of high and low tide, and the amount of vertical movement, anywhere on Earth.

There are similar tidal effects (large) on the Earth's atmosphere, and (small) on the planet's land masses. Tidal effects occur on other bodies in the solar system, and beyond. It is thought that Jupiter's moon Io, whose surface is mostly sulphur and which has numerous active volcanoes, is heated by being 'squeezed' repeatedly by tidal effects from Jupiter.

Another effect of the Moon on the Earth, discovered in the mid-'90s by Jaques Laskar, is to stabilize the Earth's axis. The Earth spins like a top, and at any given moment there is a line running through the centre of the Earth around which everything else rotates. This is its axis. The Earth's axis is tilted relative to the plane in which the Earth orbits the Sun, and this tilt is what causes the seasons. Sometimes the north pole is tilted closer to the sun than the south pole is, and six months later it's the other way round. When the northern end of the axis is tilted towards the Sun, more sunlight falls on the northern half of the planet than on the southern half, so the north gets summer and the south gets winter. Six months later, when the axis points the other way relative to the sun, the reverse applies.

Over longer periods of time, the axis changes direction. Just as a top wobbles when it spins, so does the Earth, and over 26,000 years its axis completes one full circle of wobble. At every stage, however, the axis is tilted at the same angle (23°) away from the perpendicular to the orbital plane. This motion is called precession, and it has a small effect on the timing of the seasons – they slowly shift by a total of one year in 26,000. Harmless, basically. However, the axes of most other planets do something far more drastic: they change their angle to the orbital plane. Mars, for example, probably changes this angle by 90° over a period of 10–20 million years. This has a dramatic effect on climate.

Suppose that a planet's axis is at right angles to the orbital plane. Then there are no seasonal variations at all, but everywhere except the poles there is a day/night cycle, with equal amounts of day and night. Now tilt the axis a little: seasonal variations appear, and the days are longer in summer and shorter in winter. Suppose that the axis tilts 90°, so that at some instant the north pole, say, points directly at the sun. Half a year later, the south pole points at the Sun. At either pole, there is a 'day' of half a year followed by a 'night' of half a year. The seasons coincide with the day/night cycle. Regions of the planet bake in high heat for half a year, then freeze for the other half. Although life *can* survive in such circumstances, it may be harder for it to get going in the first place, and it may be more vulnerable to extremes of climate, vulcanism, or meterorite impacts.

The Earth's axis can change its angle of tilt over very long periods of time, much longer than the 26,000 year cycle of precession, but even over hundreds of millions of years the angle doesn't change *much*. Why? Because, as Laskar discovered when he did the calculations, the Moon helps keep the Earth's axis steady. So it is at least conceivable that life on Earth owes quite a lot to the calming influence of its sister world, however much it may madden us individually.

A third influence of the Moon was discovered in 1998: a clear association between tides and the rate of growth of trees. Ernst Zürcher and Maria-Giulia Cantiani measured the diameters of young spruce trees grown in containers kept at constant light levels. Over periods of several days the diameters changed in step with the tides. The scientists interpret this as an effect of the Moon's gravity on the transport of water within the tree. It can't be variations in moonlight, which would perhaps affect photosynthesis, because the trees were grown in darkness. But the effect may be similar to one that occurs with creatures that live on the seashore. Because they evolved to live there, they have to respond to the tides, and evolution sometimes achieves this by creating an internal dynamic that runs in step with the tides. If you remove the creatures to the laboratory, this internal dynamic makes them continue to 'follow' the tides.

The Moon has been important in another way. The Babylonians and Greeks knew that the Moon is a sphere; the phases are obvious, and there is also a slight wobble which means that, over time, humans see rather more than one half of the Moon's surface. There it was, hanging in the sky – a big ball, *not* a disc like the sun, and a hint that perhaps 'big balls in space' is a much better way of thinking about the Earth and its neighbours than 'lights in the sky'.

All this is a long way from lance-constable Angua – even a long way from the female menstrual cycle. But it shows how much we are creatures of the universe. Things Up There really do affect us Down Here, every day of our lives.

TWENTY-ONE

THE LIGHT YOU SEE THE DARK BY

 THERE WAS NO DARK. This came as such a shock to Ponder Stibbons that he made HEX look again. There had to be Dark, surely? Otherwise, what was there for the light to show up against?

Eventually, he reported this lack to the other wizards.

'There should be lots of Dark and there isn't,' he said flatly. 'There's just Light and … no light. And it's a pretty strange light, too.'

'In what way?' said the Archchancellor.

'Well, sir, as you know,* there's ordinary light, which travels at about the same speed as sound …'

'That's right. You've only got to watch shadows across a landscape to realize that.'

'Quite, sir … and then there's meta–light, which doesn't really travel at all because it is already everywhere.'

'Otherwise we wouldn't even be able to see darkness,' said the Senior Wrangler.

'Exactly. But the Project universe has just got the one sort of light. HEX thinks it moves at hundreds of thousands of miles a second.'

'What use is that?'

'Er … in this universe, that's as fast as you can go.'

'That's nonsense, because –' Ridcully began, but Ponder held up a hand. He had not been looking forward to this one.

'Please, Archchancellor. It's doing the best it can. Just trust me on this one. Please? Yes, I can see all the reasons why it's impossible. But, in there, it seems to work. HEX has written pages of stuff

---

* A phrase meaning 'I'm not sure you know this.'

about it, if anyone's interested. Just don't ask me about any of it. Please, gentlemen? It's all supposed to be logical but you'll find your brain squeaking around until the ends point out of your ears.'

He placed his hands together and tried to look wise.

'It really *is* almost as if the Project is aping the *real* universe –'

'Ook.'

'I beg your pardon,' said Ponder. 'A figure of speech.'

The Librarian nodded at him and knuckled his way across the floor. The wizards watched him carefully.

'You really believe that *that* thing,' said the Dean, pointing, 'with its moon-hating water and worlds that go around suns –'

'As far as I can see from this,' interrupted the Senior Wrangler, who'd been reading HEX's write-out on the more complex physics of the Project, 'if you were travelling in a cart at the speed of light, and threw a ball ahead of you ...' he turned over the page, read on silently for a moment, creased his brows, turned the page over to see if any enlightenment was to be found on the other side, and went on '... your twin brother would ... be fifty years older than you when you got home ... I think.'

'Twins are the same age,' said the Dean, coldly. 'That's *why* they are twins.'

'Look at the world we're working on, sir,' said Ponder. 'It could be thought of as two turtle shells tied together. It's got no top and bottom but if you think of it as two worlds, bent around, with one sun and moon doing the work of two ... it's similar.'

He fried in their gaze.

'In a way, anyway,' he said.

Unnoticed by the others, the Bursar picked up the write-out on the physics of the Roundworld universe. After making himself a paper hat out of the title page, he began to read ...

TWENTY-TWO

THINGS THAT AREN'T

LIGHT HAS A SPEED — SO WHY NOT DARK?

It's a reasonable question. Let's see where it leads.

In the 1960s a biological supply company advertised a device for scientists who used microscopes. In order to see things under a microscope, it's often a good idea to make a very thin slice of whatever it is you're going to look at. Then you put the slice on a glass slide, stick it under the microscope lens, and peer in at the other end to see what it looks like. How do you make the slice? Not like slicing bread. The thing you want to cut — let's assume it's a piece of liver for the sake of argument — is too floppy to be sliced on its own.

Come to think of it, so is a lot of bread.

You have to hold the liver firmly while you're cutting it, so you embed it in a block of wax. Then you use a gadget called a microtome, something like a miniature bacon-slicer, to cut off a series of very thin slices. You drop them on the surface of warm water, stick some on to a microscope slide, dissolve away the wax, and prepare the slide for viewing. Simple ...

But the device that the company was selling wasn't a microtome: it was something to keep the wax block cool while the microtome was slicing it, so that the heat generated by the friction would not make the wax difficult to slice and damage delicate details of the specimen.

Their solution to this problem was a large concave (dish-shaped) mirror. You were supposed to build a little pile of ice cubes and 'focus the cold' on to your specimen.

Perhaps you don't see anything remarkable here. In that case you probably speak of the 'spread of ignorance', and draw the curtains in the evening to 'keep the cold out' — and the darkness.*

---

* And if so: congratulations! You are a human being, thinking narratively.

In Discworld, such things make sense. Lots of things are real in Discworld while being mere abstractions in ours. Death, for example. And Dark. On Discworld you can worry about the speed of Dark, and how it can get out of the way of the light that is ploughing into it at 600 mph.* In our world such a concept is called a 'privative' – an absence of something. And in our world, privatives don't have their own existence. Knowledge does exist, but ignorance doesn't; heat and light exist, but cold and darkness don't. Not as *things*.

We can see the Archchancellor looking puzzled, and we realize that here is something that runs quite deep in the human psyche. Yes, you can freeze to death, and 'cold' is a good word for describing the absence of heat. Without privatives, we would end up talking like the pod people from the Planet Zog. But we run into trouble, though, if we forget that we're using them as an easy shorthand.

In our world there are plenty of borderline cases. Is 'drunk' or 'sober' the privative? In Discworld you can get 'knurd', which is as far on the other side of sober as drunk is on the inebriated side,† but on planet Earth there's no such thing. By and large, we think we know which member of such a pairing has an existence, and which is merely an absence. (We vote for 'sober' as the privative. It is the absence of drink, and – usually – the normal state of a person.‡ In fact that normal state is only *called* sobriety when the subject of drink is at hand. There's nothing strange about this. 'Cold' is the normal state of the universe, after all, even though as a *thing* it does not exist. Er … we're not going to get past you on this one, are we, Archchancellor?)

---

* Light on the Disc travels at about the same speed as sound. This does not appear to cause problems.

† And a terrible thing it is, akin to a state of horrible depression. Hence the affliction of Captain Vimes in *Guards! Guards!* who needs a couple of drinks simply to become sober.

‡ Well … most people.

Thinking is required if our language isn't to fool us. However, as 'focusing the cold' shows, we sometimes don't stop to think.

We've done it before. At the start of the book, we mentioned phlogiston, considered by early chemists to be the substance that made things burn. It must do: you could see the phlogiston *coming out* as flames, for goodness' sake. Gradually, however, clues that supported the opposite view accumulated. Things weigh *more* after they've burned than they did before, for instance, so phlogiston seemed to have negative weight. You may think this is wrong, incidentally; surely the ash left by a burnt log weighs a lot less than the log, otherwise nobody would bother having bonfires? But a lot of that log goes up in smoke, and the smoke weighs quite a bit; it rises not because it's lighter than air but because it's hot. And even if it *were* lighter than air, air has weight, too. And as well as the smoke, there's steam, and all sorts of other junk. If you burn a lump of wood, and collect all the liquids, gases, and solids that result, the final total weighs more than the wood.

Where does the extra weight come from? Well, if you take the trouble to weigh the *air* that surrounds the burning wood, you'll find that it ends up *lighter* than it was. (It's not so easy to do both of these weighings while keeping track of what came from where – think about it. But the chemists found ways to achieve this.) So it looks as if something gets taken out of the air, and once you're realized that's what's going on, it's not hard to find out what it is. Of course, it's oxygen. Burnt wood gains oxygen, it doesn't lose phlogiston.

This all makes far more sense, and it also explains why phlogiston wasn't such a silly idea. Negative oxygen, oxygen that ought to be present but isn't, behaves just as nicely as positive oxygen in all the balancing equations that chemists used to check the validity of their theories. So much phlogiston moving from A to B has exactly the same effect on observations as the same amount of oxygen moving from B to A. So phlogiston behaved just like a real thing – with that embarrassing exception that when your measurements became accurate enough to detect the tiny amounts involved, phlogiston weighed less than nothing. Phlogiston was a privative.

A difficult but stubborn feature of human thinking is involved in all this: it's known as 'reifying': making real. Imagining that because we have a word for something, then there must exist a 'thing' that corresponds to the word. What about 'bravery' and 'cowardice'? Or 'tunnel'? Indeed, what about 'hole'?

Many scientific concepts refer to things that are not real in the everyday sense that they correspond to *objects*. For instance, 'gravity' sounds like an explanation of planetary motion, and you vaguely wonder what it would look like if you found some, but actually it is only a word for an inverse square law attractive relationship. Or more recently, thanks to Einstein, for a tendency of objects not to move in straight lines, which we can reify as 'curved space'.

For that matter, what about 'space'? Is that a thing, or an absence?

'Debt' and 'overdraft' are very familiar privatives, and the thinking problems they cause are quite difficult. After all, your overdraft pays your bank manager's salary, doesn't it? So how can it fail to be real? Today's derivatives market buys and sells debts and promises *as if they were real* – and it reifies them as words and numbers on pieces of paper, or digits in a computer's memory. The more you think about it, the more amazing the everyday world of human beings becomes: most of it doesn't actually exist at all.

Some years ago, at a science-fiction convention held in The Hague, four writers who made lots of money from their books sat in front of an audience of mostly impecunious fans to explain how they'd made huge income from their books (as if any of them really knew). Each of them said that 'money isn't important', and the fans became quite rude at this perfectly accurate statement. It was necessary to point out that money is like air or love – unimportant if you've got enough of it, but desperately important if you haven't.* Dickens recognized this: in *David Copperfield* Mr Micawber remarks 'Annual income twenty pounds, annual expenditure nineteen nineteen six, result happiness. Annual income twenty pounds, annual expenditure twenty pounds ought and six, result misery.'

---

* 'Desperate' is another privative – it means 'no hope'.

There's no symmetry between having money and not having it – but the discussion had gone off the rails because everyone had assumed that there was, so that 'having money' was the opposite of 'having no money'. If you must find an opposite, then 'having money' is opposite to 'being in debt'. In that case, 'rich' is like 'knurd'. In any event, making the comparison between money, love, and air lowered the debating temperature considerably. Air isn't important if you've got it, only if you haven't; the same goes for money.

Vacuum is an interesting privative. Cut-me-own-Throat Dibbler could sell vacuum-on-a-stick. Vacuum in the right place is *valuable*.

Many people on Earth sell *cold*-on-a-stick.

Discworld does a marvellous job of revealing the woolly thinking behind our assumptions about absence, because in Discworld privatives really do exist. The dark/light joke in Discworld is silly enough that everyone gets the point – we hope. Other Discworld uses of privatives, however, are more subtle. The most dramatic, of course, is Death, many people's favourite Discworld character, who SPEAKS IN CAPITAL LETTERS. Death is a seven-foot-tall skeleton, with tiny points of light in his eye sockets. He carries a scythe with a blade so thin that it's transparent, and he has a flying horse called Binky. When Death appears to Olerve, king of Sto Lat, in *Mort*, it takes the king a few moments to catch up on current events:

'Who the hell are you?' said the king. 'What are you doing here?

Eh? Guards! I deman–'

The insistent message from his eyes finally battered through to his brain. Mort* was impressed. King Olerve had held on to his throne for many years and, even when dead, knew how to behave.

'Oh,' he said. 'I see. I didn't expect to see you so soon.'

YOUR MAJESTY, said Death, bowing, FEW DO.

---

* Death's apprentice – well, he'd have to train a successor. Not in case he *dies*: so he can retire. Which he does (temporarily) in *Reaper Man*.

The king looked around. It was quiet and dim in this shadow world, but outside there seemed to be a lot of excitement.

'That's me down there, is it?'

I'M AFRAID SO, SIRE.

'Clean job. Crossbow, was it?'

Our earthly fears about death have led to some of our strangest reifications. Inventing the concept 'death' is giving a name to a process – dying – as if it's a 'thing'. Then, of course, we endow the thing with a whole suite of properties, whose care is known only to the priests. That thing turns up in many guises. It may appear as the 'soul', a thing that must leave the body when it turns it from a live body into a dead one. It is curious that the strongest believers in the soul tend to be people who denigrate material things; yet they then turn their own philosophy on its head by insisting that when an evident *process* – life – comes to an end, there has to be a *thing* that continues. No. When a process stops, it's no longer 'there'. When you stop beating an egg, there isn't some pseudo-material essence-of-eggbeater that passes on to something else. You just aren't turning the handle any more.

Another 'thing' that arises from the assumption that death exists is whatever must be instituted in the egg/embryo/foetus in order to turn it into a proper human being, who can die when required. Note that in human myth and Discworld reality it is the soulless ones, vampires and their ilk, who cannot die. Long before ancient Egypt and the death-god Anubis, priests have made capital out of this verbal confusion. On Discworld, it's entirely proper to have 'unreal' things, like Dark, or like the Tooth Fairy in *Hogfather*, which play their part in the plot.* But it's a very strange idea indeed on planet Earth.

Yet it may be part of some process that makes us human beings. As Death points out in *Hogfather*, humans seem to need to project

---

* Indeed, it is a 'fundamental constant' of the Discworld universe that things exist because they're believed in.

a kind of interior decoration on to the universe, so that they spend much of the time in a world of their own making. We seem – at least, at the moment – to need these things. Concepts like gods, truth* and soul appear to exist only in so far as humans consider them to do so (although elephants are known to get uneasy and puzzled upon finding elephant bones in the wild – whether this is because of some dim concept of the Big Savannah In The Sky or merely because it's manifestly not a good idea to stay in a place where elephants get killed is unknown). But they work some magic for us. They add narrativium to our culture. They bring pain, hope, despair, and comfort. They wind up our elastic. Good or bad, they've made us into people.

We wonder if the users thought that that cold-focusing mirror worked some magic for them. We can think of several ways in which it might appear to. And some very clever friends of ours are persuaded that souls might exist, too. Nearly everything is a process on some level. To a physicist, matter is a process carried out by a quantum wave function. And quantum wave functions exist only when the person you're arguing with asserts that they don't – so maybe souls exist in the same way.

In this area, we have to admit the science doesn't know everything. Science is *based* on not knowing everything. But it does know some things.

---

* 'Truth' is a privative in the same way that 'sober' is – until you invent lies, you don't know what the truth is. Nature appears to, otherwise animals would not have invested so much effort on very effective camouflage.

TWENTY-THREE
NO POSSIBILITY
OF LIFE

 IT WAS DIFFICULT EATING SANDWICHES that you couldn't see. Rincewind was aware that back in the real world the Librarian was handing them to him, and he had to take it on trust that they were going to be cheese and chutney. As it turned out, he detected a hint of banana, too.

The wizards were shocked. It's terrible to find that you can't do what you like with your own universe.

'So we can't just magic life into the Project?' said the Dean.

'I'm afraid not, sir,' said Ponder. 'We have quite a lot of control over things, but only in a very subtle way. I *have* gone into this.'

'I don't call moving huge worlds very subtle,' said the Dean.

'In Project terms, even moving the moon into place took a hundred thousand years,' said Ponder. 'Time prefers to move faster in there. It's amazing what you can move if you give it a little push for that long.'

'But we've done so many things –'

'Just moved things around, sir.'

'Seems a shame to have made a world and there's no one to live on it,' said the Senior Wrangler.

'When I was small, I had a model farmyard,' said the Bursar, looking up from his reading.

'Thank you, Bursar. Very interesting,' said the Archchancellor. 'All right, let's play by the rules. What do you have to move around to get people?'

'Well ... bits of other people, my father told me,' said the Dean.

'Bad taste there, Dean.'

'Many religions start with dust,' said the Senior Wrangler. 'And then you bring it alive in some way.'

'That's pretty hard even with magic,' said the Archchancellor. 'And we can't use magic.'

'Up in Nothingfjord they believe that all life was created when the god Noddi cut off his ... unmentionables and hurled them at the sun, who was his father,' said the Senior Wrangler.

'What, you mean his ... underwear?' said the Lecturer in Recent Runes, who could be a bit slow.

'First of all we can't physically exist inside the Project, secondly that sort of thing is unhygienic, and thirdly I doubt very much if you'll find a volunteer,' said the Archchancellor sharply. 'Anyway, we're men of magic. *That* is superstition.'

'Can we make weather, then?' said the Dean.

'I think HEX can let us do that,' said Ponder. 'Weather is only pushing stuff around.'

'So we can aim lightning at anyone we don't like?'

'But there isn't anyone on the world, whether we like them or not,' said Ponder wearily. 'That's the *point*.'

'And while the Dean can make enemies anywhere, I think that, ah, *Round*world would test even his powers,' said Ridcully.

'Thank you, Archchancellor.'

'Happy to oblige, Dean.'

HEX's keyboard clattered. The quill pen began to write.

It began:

+++ I Don't Think You Are Going To Believe This +++

Thunderstorms tore the air apart, far out to sea.

The air blinked. The storm was gone. The shoreline looked different.

'Hey, what happened?' said Rincewind.

'Everything all right?' said Ponder Stibbons in his ear.

'What happened just then?'

'We've moved you forward in time a little,' said Ponder. The tone of his voice suggested that he dreaded being asked why.

'Why?' said Rincewind.

'You'll laugh when I tell you this ...'

'Oh, good. I like a laugh.'

'HEX says he's detecting life all round you. Can you see anything?'

Rincewind looked around warily. The sea was sucking at the shore, which had a bit of sand on it now. Scum rolled in the waves.

'No,' he said.

'Good. You see, there *can't* be any life where you are,' Ponder went on.

'Where *am* I exactly?'

'Er ... a sort of magical world with no one in it but yourself.'

'Oh, you mean the sort everyone lives in,' said Rincewind bitterly. He glanced at the sea again, just in case.

'But if you wouldn't mind having a look ...' Ponder went on.

'For this life that can't possibly exist?'

'Well, you are the Professor of Cruel and Unusual Geography.'

'It's the cruel and unusual geography that's bothering me,' said Rincewind. 'Incidentally, have you looked at the sea lately? It's blue.'

'Well? The sea *is* blue.'

'Really?'

The omniscope was once again the centre of attention.

'Everyone knows the sea is blue,' said the Dean. 'Ask anyone.'

'That's right,' said Ridcully. 'However, while everyone *knows* the sea is blue, what everyone usually sees is a sea that's grey or dark green. Not *this* colour. This is virulent!'

'I'd say turquoise,' said the Senior Wrangler.

'I used to have a shirt that colour,' said the Bursar.

'I thought it might be copper salts in the water,' said Ponder Stibbons. 'But it isn't.'

The Archchancellor picked up HEX's latest write-out. It read:

+++ Out Of Cheese Error +++

'Not helpful,' he muttered.

'Thank goodness he's still operating the Project,' said Ponder, joining him. 'I think he's got confused.'

'It's not his job to be confused,' said Ridcully. 'We don't need a machine for being confused. We're entirely capable of confusin' ourselves. It is a human achievement, confusion, and right at this minute I feel I am winning a prize. *You*, Mister Stibbons, said there was no possibility of life turnin' up inside the Project.'

Ponder waved his hands frantically. 'There's no way that it can! Life isn't like rocks and water. Life is special!'

'The breath of gods, that sort of thing?' said Ridcully.

'Not gods as *such*, obviously, but –'

'I suppose from the point of view of rocks, rocks are special,' said Ridcully, still reading HEX's output.

'No, sir. Rocks don't have a point of view.'

Rincewind lifted up a shard of rock, very carefully, ready to drop it immediately at the merest suggestion of tooth or claw.

'This is silly,' he said. 'There's nothing here.'

'Nothing?' said Ponder, inside the helmet.

'Some of the rocks have got all kind of yuk on them, if that's your idea of a good time.'

'Yuk?'

'You know … gunge.'

'HEX seems to be suggesting now that whatever is showing up is, and is not, life,' said Ponder, a man whose interest in slime was limited.

'That's very cheering.'

'There seems to be a particular concentration not far from you … we're just going to move you so that you can have a look at it …'

Rincewind's head swam. A moment later, the rest of his body wanted to join it. He was underwater.

'Don't worry,' said Ponder, 'because although you're at a very great depth, the pressure can't possibly hurt you.'

'Good.'

'And the boiling water should feel merely tepid.'

'Fine.'

'And the terrible upflow of poisonous minerals can't harm you because of course you're not really there.'

'So, all in all, I'm laughing,' said Rincewind gloomily, peering at the dim glow ahead of him.

'It's gods, definitely,' said the Archchancellor. 'Gods have turned up while our back was turned. There can be no other explanation.'

'Then they seem rather unambitious,' sniffed the Senior Wrangler. 'I mean, you'd expect humans, wouldn't you? Not ... blobs you can't see. They're not going to bow down and worship anyone, are they?'

'Not where they are,' said Ridcully. 'The planet's full of cracks! You shouldn't get fire *under* water. That's against nature!'

'Everywhere you look, little blobs,' said the Senior Wrangler. 'Everywhere.'

'Blobs,' said the Lecturer in Recent Runes. 'Can they pray? Can they build temples? Can they wage holy war on less enlightened blobs?'

Ponder shook his head sadly. Hex's results were quite clear. Nothing solid could cross the barrier into Roundworld. It was possible, with enough thaumic effort, to exert tiny pressures, but that was all. Of course, you could speculate that thought might get in there, but if that was the case the wizards were thinking some very dull thoughts indeed. 'Blobs' wasn't really a good word for what were currently floating in the warm seas and dribbling over the rocks. It had far too many overtones of feverish gaiety and excitement.

'They're not even moving,' said Ridcully. 'Just bobbing about.'

'Blobbing about, haha,' said the Senior Wrangler.

'Could we ... help them in some way?' said the Lecturer in Recent Runes. 'You know ... to become better blobs? I fear we have some responsibility.'

'They may be as good as blobs get,' said Ridcully. 'What's up with that Rincewind fellow?'

They turned. In its circle of smoke, the suited figure was making frantic running motions.

'Do you think, on reflection, that it might not have good idea to miniaturize his image in Roundworld?' said Ridcully.

'It was the only way we could get him into that little rock pool Hex wanted us to look at, sir,' said Ponder. 'He doesn't *have* to be any particular size. Size is relative.'

'Is that why he keeps calling out for his mother?'

Ponder went over to the circle and rubbed out a few important runes. Rincewind collapsed on the floor.

'What idiot put me in *there*?' he said. 'Ye gods, it's awful! The size of some of those things!'

'They're actually tiny,' said Ponder, helping him up.

'Not when you are smaller than them!'

'My dear chap, they can't possibly *hurt* you. You have nothing to fear but fear itself.'

'Oh, is that so? What help is that? You think that makes it better? Well, let me tell you, some of that fear can be pretty big and nasty –'

'Calm down, calm *down*.'

'Next time I want to be big, understand?'

'Did they try to communicate with you in any way?'

'They just flailed away with great big whiskers! It was worse than watching wizards arguing!'

'Yes, I doubt if they are very intelligent.'

'Well, nor are the rock pool creatures.'

'I'm just wondering,' said Ponder, wishing he had a beard to stroke thoughtfully, 'if perhaps they might … improve with keeping …'

TWENTY-FOUR

DESPITE WHICH ...

THAT BLUE IN THE ROUNDWORLD SEA isn't a chemical – well, not in the usual '*simple* chemical' sense of the word. It's a mass of bacteria, called cyanobacteria. Another name for them is 'blue-green algae', which is wonderfully confusing. *Modern* so-called blue-green algae are usually red or brown, but the ancient ones probably *were* blue-green. And blue-green algae are really bacteria, whereas most other algae have cells with a nucleus and so are not bacteria. The blue-green colour comes from chlorophyll, but of a different kind from that in plants, together with yellow-orange chemicals called carotenoids.

Bacteria appeared on Earth at least 3.5 billion years ago, only a few hundred million years after the Earth cooled to the point at which living creatures could survive on it. We know this because of strange layered structures found in sedimentary rocks The layers can be flat and bumpy, they can form huge branched pillars, or they can be highly convoluted like the leaves in a cabbage. Some deposits are half a mile thick and spread for hundreds of miles. Most date from 2 billion years ago, but those from Warrawoona in Australia are 3.5 billion years old.

To begin with, nobody knew what these deposits were. In the 1950s and 1960s they were revealed as traces of communities of bacteria, especially cyanobacteria.

Cyanobacteria collect together in shallow water to form huge, floating mats, like felt. They secrete a sticky gel as protection against ultraviolet light, and this causes sediment to stick to the mats. When the layer of sediment gets so thick that it blocks out the light, the bacteria form a new layer, and so on. When the layers fossilize they turn into stromatolites, which look rather like big cushions.

The wizards haven't been expecting life. Roundworld runs on

rules, but life doesn't – or so they think. The wizards see a sharp discontinuity between life and non-life. This is the problem of expecting *becomings* to have boundaries – of imagining that it ought to be easy to class all objects into either the category 'alive' or the category 'dead'. But that's not possible, even ignoring the flow of time, in which 'alive' can *become* 'dead' – and vice versa. A 'dead' leaf is no longer part of a living tree, but it may well have a few revivable cells.

Mitochondria, now the part of a cell that generates its chemical energy, once used to be independent organisms. Is a virus alive? Without a bacterial host it can't reproduce – but neither can DNA copy itself without a cell's chemical machinery.

We used to build 'simple' chemical models of living processes, in the hope that a sufficiently complex network of chemistry could 'take off' – become self-referential, self-copying – by itself. There was the concept of the 'primal soup', lots of simple chemicals dissolved in the oceans, bumping into each other at random, and just occasionally forming something more complicated. It turns out that this isn't quite the way to do it. You don't have to work hard to make real-world chemistry complex: that's the default. It's *easy* to make complicated chemicals. The world is full of them. The problem is to keep that complexity organized.

What counts as life? Every biologist used to have to learn a list of properties: ability to reproduce, sensitivity to its environment, utilization of energy, and the like. We have moved on. 'Autopoeisis' – the ability to make chemicals and structures related to one's own reproduction – is not a bad definition, except that modern life has evolved away from those early necessities. Today's biologists prefer to sidestep the issue and define life as a property of the DNA molecule, but this begs the deeper question of life as a general *type* of process. It may be that we're now defining life in the same way that 'science fiction' is defined – it's what we're pointing at when we use the term.*

The idea that life could somehow be self-starting is still controversial to many people. Nevertheless, it turns out that finding plausible routes to life is easy. There must be at least thirty of them.

It's hard to decide which, if any, was the actual route taken, because later lifeforms have destroyed nearly all the evidence. This may not matter much: if life hadn't taken the route that it did, it could easily have taken one of the others, or one of the hundred we haven't thought of yet.

One possible route from the inorganic world to life, suggested by Graham Cairns-Smith, is clay. Clay can form complicated microscopic structures, and it often 'copies' an existing structure by adding an extra layer to it, which then falls off and becomes the starting point of a new structure. Carbon compounds can stick on to clay surfaces, where they can act as catalysts for the formation of complex molecules of the kind we see in living creatures – proteins, even DNA itself. So today's organisms may have hitched an evolutionary ride on clay.

An alternative is Günter Wächtershäuser's suggestion that pyrite, a compound of iron and sulphur, could have provided an energy source suitable for bacteria. Even today we find bacteria miles underground, and near volcanic vents at the bottom of the oceans, which power themselves by iron/sulphur reactions. These are the source of the 'upflow of poisonous minerals' noticed by Rincewind. It's entirely conceivable that life started in similar environments.

A potential problem with volcanic vents, though, is that every so often they get blocked, and another one breaks out somewhere else. How could the organisms get themselves safely across the intervening cold water? In 1988 Kevin Speer realized that the Earth's

---

* Everyone knows what science fiction is – until you start asking questions like 'Is a book set five years in the future *automatically* SF? Is it SF just because it's set on another world, or is it simply fantasy with nuts and bolts on the outside? Is it SF if the author thinks it isn't? Does it have to be set in the future? Does the presence of Doug McClure mean that a movie is SF, or merely that the men-in-rubber-monster-suits quotient is going to be high?' One of the best SF books ever written was the late Roy Lewis's *The Evolution Man;* there is no technology in it more sophisticated than a bow, it's set in the far past, the characters are barely more than ape-men ... but it is science fiction, nonetheless.

rotation causes the rising plumes of hot water from vents to spin, forming a kind of underwater hot tornado that moves through the deep ocean. Organisms could hitch a ride on these. Some might make it to another vent. Many would not, but that doesn't matter – all that would be required would be *enough* survivors.

It is interesting to note that back in the Cretaceous, when the seas were a lot warmer than now, these hot plumes could even have risen to the ocean's surface, where they may have caused 'hyper-canes' – like hurricanes but with a windspeed close to that of sound. These would have caused major climatic upheavals on a planet which, as we shall see, it not the moderately peaceful place we tend to believe it is.

Bacteria belong to the grade of organisms known as prokaryotes. They are often said to be 'single-celled', but many single-celled creatures are far more complex and very different from bacteria. Bacteria are not true cells, but something simpler; they have no cell wall and no nucleus. True cells, and creatures both single-celled and many-celled, came later, and are called eukaryotes. They probably arose when several different prokaryotes joined forces to their mutual benefit – a trick known as symbiosis. The first fossil eukary-otes are singe-celled, like amoebas, and appear about 2 billion years ago. The first fossils of many-celled creatures are algae from 1 billion years ago ... maybe even as old as 1.8 billion years.

This was the story as scientists understood it up until 1998: animals like arthropods and other complex beasts came into being a mere 600 million years ago, and that until about 540 million years ago the only creatures were very strange indeed – quite unlike most of what's around today.

These creatures are known as Ediacarans, after a place in Australia where the first fossils were found.* They could grow to half a metre or more, but as far as can be told from the fossil record, seem not to have had any internal organs or external orifices like a

---

* They were fortunate, given the names of some places in Australia, that they ended up merely *sounding* like a minor *Star Trek* species.

mouth or an anus (they may have survived by digesting symbiotic bacteria in their selves, or by some other process we can only guess at). Some were flattened, and clustered together in quilts. We have no idea whether the Ediacarans were our distant ancestors, or whether they were a dead end, a lifestyle doomed to failure. No matter: they were around then, and as far as anyone knew, not much else was. There are hints of fossil wormcasts, though, and some very recent fossils look like ... but we're getting ahead of the story. The point is that nearly all Ediacaran life was apparently unrelated to what came later.

About 540 million years ago the Pre-Cambrian Ediacarans were succeeded by the creatures of the Cambrian era. For the first ten million years, these beasties were also pretty weird, leaving behind fragments of spines and spikes which presumably are the remains of prototype skeletons that hadn't yet joined up. At that point, nature suddenly learned how to do joined-up skeletons, and much else: this was the time known as the Cambrian Explosion. Twenty million years later virtually every body-plan found in modern animals was already in existence: everything afterwards was mere tinkering.

The real innovation of the Cambrian Explosion, though, was less obvious than joined-up skeletons or tusks or shells or limbs. It was a *new kind of body plan*. Diploblasts were overtaken by triploblasts ...

Sorry, Archchancellor. We mean that creatures began to put another layer between themselves and the universe. Ediacarans and modern jellyfish are diploblasts – two-layered creatures. They have an inside and an outside, like a thick paper bag. Three-layered creatures like us and practically everything else around are called triploblasts. We have an inner, an outer, and a *within*.

The *within* was the big leap forward, or at least the big slither. *Within* you can put the things you need to protect, like internal organs. In one sense, you are not part of the environment any more there is a *you* as well. And, like someone who now has a piece of property of their very own, you can begin to make improvements.

This is a lie-to-children, but as lies go it is a good one.

Triploblasts played a crucial role in evolution, precisely because they *did* have internal organs, and in particular they could ingest food and excrete it. Their excreta became a major resource for other creatures; to get an interestingly complicated world, it is vitally important that shit happens.

But where did all those triploblasts come from? Were they an offshoot of the Ediacarans? Or did they come from something else that didn't leave fossils?

It's hard to see how they could have come from Ediacarans. Yes, an extra layer of tissue might have appeared, but as well as that extra layer you need a lot of organization to exploit it. That organization has to come from somewhere. Moreover, there were these occasional tantalizing traces of what might have been pre-Cambrian triploblasts – fossils not of worms, which would have clinched it, but of things that might have been trails made by worms in wet mud.

And then again, might not.

In February 1998, we found out.

The discovery depended upon where – and in this case how – you look for fossils. One way for fossils to form is by petrification. There is a poorly known type of petrification that can happen *very* fast – within a few days. The soft parts of a dead organism are replaced by calcium phosphate. Unfortunately for palaeontologists, this process works only for organisms that are about a tenth of an inch (2 mm) long. Still, some interesting things are that tiny. From about 1975 onwards scientists found wonderfully preserved specimens of tiny ancient arthropods – creatures like centipedes with many segments. In 1994 they found fossilized balls of cells from embryos – early stages in the development of an organism – and it is thought that these come from embryonic triploblasts. However, all of these creatures must have come *after* the Ediacarans. But in 1998 Shuhai Xiao, Yun Zhang, and Andrew Knoll discovered fossilized embryos in Chinese rock that is 570 million years old – smack in the middle of the Ediacaran era. And those embryos were *triploblasts*.

Forty million years before the Cambrian explosion, there were triploblasts on Earth, living right alongside those enigmatic

Ediacarans.

We are triploblasts. Somewhere in the pre-Cambrian, surrounded by mouthless, organless Ediacarans, we came into our inheritance.

It used to be thought that life was a delicate, highly unusual phenomenon: difficult to create, easy to destroy. But everywhere we look on Earth we find living creatures, often in environments that we would have expected to be impossibly hostile. It's beginning to look as if life is an extremely robust phenomenon, liable to turn up almost anywhere that's remotely suitable. What is it about life that makes it so *persistent*?

Earlier we talked about two ways to get off the Earth, a rocket and a space elevator. A rocket is a thing that gets used up, but a space elevator is a process that continues. A space elevator requires a huge initial investment, but once you've got it, going up and down is essentially free. A functioning space elevator seems to contradict all the usual rules of economics, which look at individual transactions and try to set a rational price, instead of asking whether the concept of a price might be eliminated altogether. It also seems to contradict the law of conservation of energy, the physicist's way of saying that you can't get something for nothing. But, as we've seen, you can – by exploiting the new resources that become available once you get your space elevator up and running.

There is an analogy between space elevators and life. Life seems to contradict the usual rules of chemistry and physics, especially the rule known as the second law of thermodynamics, which says that things can't spontaneously get more complicated. Life does this because, like the space elevator, it has lifted itself to a new level of operation, where it can gain access to things and processes that were previously out of the question. Reproduction, in particular, is a wonderful method of getting round the difficulties of manufacturing a really complicated thing. Just build one that manufactures more of itself. The first one may be incredibly difficult – but all the rest come with no added effort.

What is the elevator for life? Let's try to be general here, and look at the common features of all the different proposals for 'the' origin of life. The main one seems to be the novel chemistry that can occur in small volumes adjacent to active surfaces. This is a long way from today's complex organisms – it's even a long way from today's bacteria, which are distinctly more complicated than their ancient predecessors. They have to be, to survive in a more complicated world. Those active surfaces could be in underwater volcanic vents. Or hot rocks deep underground. Or they could be seashores. Imagine layers of complicated (because that's easy) but disorganized (ditto) molecular gunge on rocks which are wetted by the tides and irradiated by the sun. Anything in there that happens to produce a tiny 'space elevator' establishes a new baseline for further change. For example, photosynthesis is a space elevator in this sense. Once some bit of gunge has got it, that gunge can make use of the sun's energy instead of its own, churning out sugars in a steady stream. So perhaps 'the' origin of life was a whole series of tiny 'space elevators' that led, step by step, to organized but ever more complex chemistry.

TWENTY-FIVE

UNNATURAL SELECTION

 THE LIBRARIAN KNUCKLED SWIFTLY through the outer regions of the University's library, although terms like 'outer' were hardly relevant in a library so deeply immersed in L-space.

It is known that knowledge is power, and power is energy, and energy is matter, and matter is mass, and therefore large accumulations of knowledge distort time and space. This is why all bookshops look alike, and why all *second-hand* bookshops seem so much bigger on the inside – and why all libraries, everywhere, are connected. Only the innermost circle of librarians know this, and take care to guard the secret. Civilization would not survive for long if it was generally known that a wrong turn in the stacks would lead into the Library of Alexandria just as the invaders were looking for the matches, or that a tiny patch of floor in the reference section is shared with the library in Braseneck College where Dr Whitbury *proved* that gods cannot possibly exist, just before that rather unfortunate thunderstorm.

The Librarian was saying 'ook ook' to himself under his breath, in the same way that a slightly distracted person searches aimlessly around the room saying 'scissors, scissors' in the hope that this will cause them to re-materialize. In fact he was saying 'evolution, evolution'. He'd been sent to find a good book on it.

He had a very complicated reference card in his mouth.

The wizards of UU knew all about evolution. It was a self-evident fact. You took some wolves, and by careful unnatural selection over the generations you got dogs of all shapes and sizes. You took some sour crab-apple trees and, by means of a stepladder, a fine paintbrush and a lot of patience, you got huge juicy apples. You took some rather scruffy desert horses and, with effort and a good stock book, you got a winner. Evolution was a demonstration of narrativium in action. Things improved. Even the human race was evolving, by means of education and other benefits of civilization; it had began with rather bad-mannered people in caves, and it had now produced the Faculty of Unseen University, beyond which it was probably impossible to evolve further.

Of course, there were people who occasionally advanced more radical ideas, but they were like the people who thought the world really *was* round or that aliens were interested in the contents of their underwear.

Unnatural selection was a fact, but the wizards knew, they *knew*, that you couldn't start off with bananas and get fish.

The Librarian glanced at the card, and took a few surprising turnings. There was the occasional burst of noise on the other side of the shelves, rapidly changing as though someone was playing with handfuls of sound, and a flickering in the air. Someone talking was replaced with the absorbent silence of empty rooms was replaced with the crackling of flame and displaced by laughter ...

Eventually, after much walking and climbing, the Librarian was faced with a blank wall of books. He stepped up to them with librarianic confidence and they melted away in front of him.

He was in some sort of study. It was book-lined, although with rather fewer than the Librarian would have expected to find in such an important node of L-space. Perhaps there was just the *one* book ... and there it was, giving out L-radiation at a strength the Librarian had seldom encountered outside the seriously magical books in the locked cellars of Unseen University. It was a book and father of books, the progenitor of a whole race that would flutter down the centuries ...

It was also, unfortunately, still being written.

The author, pen still in hand, was staring at the Librarian as if he'd seen a ghost.

With the exception of his bald head and a beard that even a wizard would envy, he looked very, very much *like* the Librarian.

'My goodness ...'

'Ook?' The Librarian had not expected to be seen. The writer must have something very pertinent on his mind.

'What manner of shade are you … ?'

'Ook.'*

A hand reached out, tremulously. Feeling that something was expected of him, the Librarian reached out as well, and the tips of the fingers touched.

The author blinked.

'Tell me, then,' he said, 'is Man an ape, or is he an angel?'

The Librarian knew this one.

'Ook,' he said, which meant: ape is best, because you don't have to fly and you're allowed sex, unless you work at Unseen University, worst luck.

Then he backed away hurriedly, ooking apologetic noises about the minor error in the spacetime coordinates, and knuckled off through the interstices of L-space and grabbed the first book he found that had the word 'Evolution' in the title.

The bearded man went on to write an even more amazing book. If only he had thought to use the word *'Ascent'* there might not have been all that unpleasantness.

But, there again, perhaps not.

HEX let itself absorb more of the future … call it … *knowledge*. Words were so difficult. Everything was context. There was too much to learn. It was like trying to understand a giant machine when you didn't understand a screwdriver.

Sometimes HEX thought it was picking up fragmentary instructions. And, further away, much further away, there were little disjointed phrases in the soup of concepts which made sense but did not seem to be sensible. Some of them arrived unbidden.

Even as HEX pondered this, another one arrived and offered an opportunity to make $$$$ While You Sit On Your Butt!!!!! He considered this unlikely.

The title brought back by the Librarian was *The Young Person's Guide to Evolution*.

* 'Reddish-brown'.

The Archchancellor turned the pages carefully. They were well illustrated. The Librarian knew his wizards.

'And this is a good book on evolution?' said the Archchancellor.

'Ook.'

'Well, it makes no sense to *me*,' said the Archchancellor. 'I mean t'say, what the hell is this picture all about?'

It showed, on the left, a rather hunched-up, ape-like figure. As it crossed the page, it gradually arose and grew considerably less hairy until it was striding confidently towards the edge of the page, perhaps pleased that it had essayed this perilous journey without at any time showing its genitals.

'Looks like me when I'm getting up in the mornings,' said the Dean, who was reading over his shoulder.

'Where'd the hair go?' Ridcully demanded.

'Well, some people shave,' said the Dean.

'This is a very *strange* book,' said Ridcully, looking accusingly at the Librarian, who kept quiet because in fact he was a little worried. He rather suspected he might have altered history, or at least *a* history, and on his flight back to the safety of UU he'd seized the first book that looked as though it might be suitable for people with a very high IQ but a mental age of about ten. It had been in an empty byway, far off his usual planes of exploration, and there had been very small red chairs in it.

'Oh, I get it. This is a fairy story,' said Ridcully. 'Frogs turnin' into princes, that kind of thing. See here ... there's something like our blobs, and then these fishes, and then it's a ... a newt, and then it's a big dragony type of thing and, hah, then it's a mouse, then here's an ape, and then it's a man. This sort of thing happens all the time out in the really rural areas, you know, where some of the witches can be quite vindictive.'

'The Omnians believe something like this, you know,' said the Senior Wrangler. 'Om started off making simple things like snakes, they say, and worked his way up to Man.'

'As if life was like modelling clay?' said Ridcully, who was not a patient man with religion. 'You start out with simple things and then progress to elephants and birds which don't stand up properly when you put them down? We've *met* the God of Evolution, gentlemen … remember? Natural evolution merely improves a species. It can't *change* anything.'

His finger stabbed at the next page in the brightly coloured book.

'Gentlemen, this is merely some sort of book of magic, possibly about the Morphic Bounce Hypothesis.* Look at this.' The picture showed a very large lizard followed by a big red arrow, followed by a bird. 'Lizards don't turn into birds. If they did, why have we still got lizards? Things can't decide for *themselves* what shape they're going to be. Ain't that so, Bursar?'

The Bursar nodded happily. He was halfway through Hex's write-out of the theoretical physics of the project universe and, so far, had understood every word. He was particular happy with the limitations of light speed. It made absolute sense.

He took a crayon and wrote in the margin: 'Assuming the universe to be a negatively curved non-Paramidean manifold – which is more or less obvious – you could deduce its topology by observ-

---

* … which had engrossed wizards for many years. The debate ran like this: it was quite easy to turn someone into a frog, and fairly easy to turn them into, say, a white mouse. Strangely, considering the basic similarity of size and shape, turning someone into an orangutan took a vast amount of power and it was only an explosion in the intense thaumic confines of the Library which had managed the trick. Turning someone into a tree was much, much harder even than that, although turning a pumpkin into a coach was so easy that even a crazy old woman with a wand could do it. Was there some kind of framework into which all this fitted?

The current hypothesis was that most Change spells unravelled the victim's morphic field down to some very basic level and then 'bounced' them back. A frog was quite simple, so they wouldn't have to bounce far. An ape, being quite human-like in many respects, would mean a very long return journey indeed. You couldn't turn someone into a tree because there was no way to get there from here, but a pumpkin could be turned into a wooden coach because it was quite close to it in vegetable space.

The wizards agreed that this all seemed to fit nicely, and was therefore true.

If William of Occam had been a wizard at Unseen University, he would have grown a beard.

ing the same galaxies in several different directions.' He thought for a moment, and added: 'Some travel will be involved.'

Of course, he was a natural mathematician, and one thing a natural mathematician wants to do is get away from actual damn sums as quickly as possible and slide into those bright sunny uplands where everything is explained by letters in a foreign alphabet, and no one shouts very much. This was even better than that. The hard-to-digest idea that there were dozens of dimensions rolled up where you couldn't see them was sheer jelly and ice cream to a man who saw *lots* of things no one else saw.

TWENTY-SIX

THE DESCENT
OF DARWIN

THE WIZARDS MET THE GOD OF EVOLUTION in *The Last Continent*. He made things the way a god ought to:

"'Amazin' piece of work," said Ridcully, emerging from the elephant. "Very good wheels. You paint these bits before assembly, do you?"'

The God of Evolution builds creatures piece by piece, like a butcher in reverse. He likes worms and snakes because they're very easy – you can roll them out like a child with modelling clay. But once the God of Evolution has *made* a species, can it change? It does on Discworld, because the God runs around making hurried adjustments ... but how does it work without such divine intervention?

All societies that have domestic animals, be they hunting dogs or edible pigs, know that living creatures can undergo gradual changes in form from one generation to the next. Human intervention, in the form of 'unnatural selection', can *breed* long thin dogs to go down holes and big fat pigs that provide more bacon per trotter.* The wizards know this, and so did the Victorians. Until the nineteenth century, though, nobody seems to have realized that a very similar process might explain the remarkable diversity of life on Earth, from bacteria to bactrians, from oranges to orangutans.

They didn't appreciate that possibility for two reasons. When you bred dogs, what you got was a different kind of dog – not a banana or a fish. And breeding animals was the purest kind of magic: if a human being *wanted* a long thin dog, and if they started from short fat ones, and if they knew how the trick worked (if, so to speak, they

---

* The quantity of bacon per trotter is on average slightly more than one quarter of the amount per head.

cast the right 'spells') then they would *get* a long thin dog. Bananas, long and thin though they might be, were *not* a good starting point. Organisms couldn't change species, and they only changed form within their *own* species because people wanted them to.

Around 1850, two people independently began to wonder whether nature might play a similar game, but on a much longer timescale and in a much grander manner – and without any sense of purpose or goal (which had been the flaw in previous musings along similar lines). They considered a self-propelled magic: 'natural' selection as opposed to selection by people. One of them was Alfred Wallace; the other – far better known today – was Charles Darwin. Darwin spent years travelling the world. From 1831 to 1836 he was hired as ship's naturalist aboard HMS *Beagle*, and his job was to observe plants and animals and note down what he saw. In a letter of 1877 he says that while on the *Beagle* he believed in 'the permanence of species', but on his return home in 1836 he began to think about the deeper meaning of what he had seen, and realized that 'many facts indicated the common descent of species'. By this he meant that species that are different now probably came from ancestors that once belonged to the *same* species. Species must be able to change. That wasn't an entirely new idea, but he also came up with an effective *mechanism* for such changes, and that *was* new.

Meanwhile Wallace was studying the flora and fauna of Brazil and the East Indies, and comparing what he saw in the two regions, and was coming to similar conclusions – and much the same explanation. By 1858 Darwin was still mulling over his ideas, contemplating a grand publication of everything he wanted to say about the subject, while Wallace was getting ready to publish a short article containing the main idea. Being a true English gentleman, Wallace warned Darwin of his intentions so that Darwin could publish something first, and Darwin rapidly penned a short paper for the Linnaean Society, followed a year later by a book, *The Origin of Species* – a big book, but still not on the majestic scale that Darwin had originally intended. Wallace's paper appeared in the same journal shortly afterwards, but both papers were officially 'presented' to the Society at the same meeting.

What was the initial reaction to these two Earth-shattering articles? In his annual report for that year, the President of the Society, Thomas Bell, wrote that 'The year has not, indeed, been marked by any of those striking discoveries which at once revolutionize, so to speak, the department of science in which they occur.' However, this perception quickly changed as the sheer enormity of Darwin's and Wallace's theory began to sink in, and they took a lot of stick from Mustrum Ridcully's spiritual brethren for daring to come up with a plausible alternative to Biblical creation. What was this epoch-making alternative? An idea so simple that everybody else had missed it. Thomas Huxley is said to have remarked, on reading *Origin:* 'How extremely stupid not to have thought of that.'

This is the idea. You don't need a human being to push animals into new forms; they can do it to themselves – more precisely: *to each other*. This was the mechanism of natural selection. Herbert Spencer, who did the important journalistic job of interpreting Darwin's theory to the masses, coined the phrase, 'survival of the fittest' to describe it. The phrase had the advantage of convincing everybody that they understood what Darwin was saying, and it had the disadvantage of convincing everybody that they understood what Darwin was saying. It was a classic lie-to-children, and it deceives many critics of evolution to this day, causing them to aim at a long-disowned target, besides giving a spurious 'scientific' background to some extremely stupid and unpleasant political theories.

Starting from an enormous range of observations of many species of plants and animals, Darwin had become convinced that organisms could change of their own accord, so much so that they could even – over *very* long periods – change so much that they gave rise to new species.

Imagine a lot of creatures of the same species. They are in competition for resources, such as food – competing with each other, and with animals of other species. Now suppose that by random chance, one or more of these animals has offspring that are better at winning the competition. Then those animals are more likely to survive for long enough to produce the next generation, and the next generation is *also* better at winning. In contrast, if one or more of these animals has offspring that are worse at winning the competition, then those animals are less likely to produce a succeeding generation – and even if they somehow do, that next generation is still worse at winning. Clearly even a tiny advantage will, over many generations, lead to a population composed almost entirely of the new high-powered winners. In fact, the effect of any advantage grows like compound interest, so it doesn't take all that long.

Natural selection sounds like a very straightforward idea, but words like 'competition' and 'win' are loaded. It's easy to get the wrong impression of just how subtle evolution must be. When a baby bird falls out of the nest and gets gobbled up by a passing cat, it is easy to see the battle for survival as being fought between bird and cat. But if *that* is the competition, then cats are clear winners – so why haven't birds evolved away altogether? Why aren't there just cats?

Because cats and birds long ago came, unwittingly, to a mutual accommodation in which *both* can survive. If birds could breed unchecked, there would soon be far too many birds for their food supply to support them. A female starling, for instance, lays about 16 eggs in her life. If they all survived, and this continued, the starling population would multiply by eight every generation – 16 babies for every two parents. Such 'exponential' growth is amazingly rapid: by the 70th generation a sphere the size of the solar system would be occupied entirely by starlings (instead of by pigeons, which appears to be its natural destiny).

The only 'growth rate' for the population that works is for each breeding pair of adult starlings to produce, on average, exactly one breeding pair of adult starlings. Replacement, but no more – and no less. Anything more than replacement, and the population explodes; anything less, and it eventually dies out. So of those 16 eggs, an average of 14 must not survive to breed. And that's where the cat comes in, along with all the other things that make it tough to be a bird, especially a young one. In a way, the cats are doing the birds a favour – collectively, though maybe not as individuals. (It depends if you're one of the two that survive to breed or the 14 that don't.)

Rather more obviously, the birds are doing the cats a favour – cat food literally drops out of the skies, manna from heaven. So what stops it getting out of hand is that if a group of greedy cats happens to evolve somewhere, they rapidly eat themselves out of existence again. The more restrained cats next door survive to breed, and quickly take over the vacated territory. So those cats that eat just enough birds to maintain their food supply will win a competition *against the greedy cats*. Cats and birds aren't *competing* because they're not playing the same game. The real competitions are between cats and other cats, and between birds and other birds. This may seem a wasteful process, but it isn't. A female starling has no trouble laying her 16 eggs. Life is reproductive – it makes reasonably close, though not exact, copies of itself, in quantity, and 'cheaply'. Evolution can easily 'try out' many different possibilities, and discard those that don't work. And that's an astonishingly effective way to home in on what *does* work.

As Huxley said, it's such an obvious idea. It caused so much trouble from religionists because it takes the gloss off one of their favourite arguments, the argument from design. Living creatures seem so perfectly put together that surely they *must* have been designed – and if so, there must have been a Designer. Darwinism made it clear that a process of random, purposeless variation trimmed by self-induced selection can achieve equally impressive results, so there can be the semblance of design without any Designer.

There are plenty of details to Darwinism that still aren't understood, as with all science, but most of the obvious ways of trying to shoot it down have been answered effectively. The classic example – still routinely trotted out by creationists and others even though Darwin himself had a pretty good answer – is the evolution of the eye. The human eye is a complex structure, and all of its components have to fit together to a high degree of accuracy, or it won't work. If we claim that such a complex structure has evolved, we must accept that it evolved gradually. It can't all have come into being at once. But if so, then at every stage along the evolutionary track the still-evolving proto-eye must offer some kind of survival advantage to the creature that possesses it. How can this happen? The question is often asked in the form 'What use is half an eye?', to which you are expected to conclude 'nothing', followed by a rapid conversion to some religion or other. 'Nothing' is a reasonable answer – but to the wrong question. There are lots of ways to get to an eye gradually that do not require it to be assembled piece by piece like a jigsaw puzzle. Evolution does not build creatures piece by piece like the God of Evolution in *The Last Continent*. Darwin himself pointed out that in creatures alive in his day you could find all kinds of light-sensitive organs – starting with patches of skin, then increasing in complexity, light-gathering power, and ability to detect fine detail, right up to structures as sophisticated as the human eye. There is a continuum of eyelike organs in the living world, and every creature gains an advantage by having its own type of light-sensing device, in comparison to similar creatures that have a slightly less effective device of a similar kind.

In 1994 Daniel Nilsson and Susanne Pelger used a computer to see what would happen to a mathematical model of a light-sensing surface if it was allowed to change in small, random, biologically feasible ways, with only those changes that improved its sensitivity to light being retained. They found that within 400,000 generations – an evolutionary blink of an eye – that flat surface gradually changed into a recognizable eye, complete with a lens. The lens even bent light differently in different places, just like our eye and unlike normal spectacle lenses. At every tiny step along the way, a creature with the improved 'eye' would be better than those with the old version.

At no stage was there ever 'half an eye'. There were just light-sensing things that got *better* at it.

Since the 1950s, we have been in possession of a new and central piece of the evolutionary jigsaw, one that Darwin would have given his right arm to know about. This is the physical – more precisely, chemical – nature of whatever it is that ensures that characteristics of organisms can change *and* be passed from one generation to the next.

You know the word: *gene*.

You know the molecule: *DNA*.

You even know how it works: DNA carries the *genome*, which is a chemical 'blueprint' for an organism. It uses the *genetic code* to turn DNA into proteins.

And, probably, a lot of what you know is lies-to-children.

Just as 'survival of the fittest' captured the imaginations of the Victorians, so 'DNA' has captured the imaginations of today's public. However, imaginations thrive best if they are left free to roam: they grow tired and feeble in captivity. Captive imaginations do breed quite effectively, because they are protected from the terrible predator known as Thought.

DNA has two striking properties, which play a significant role in the complex chemistry of life: it can encode information, and that information can be copied. (Other molecules *process* the DNA information, for example by making proteins according to recipes encoded in DNA.) From this point of view a living organism is a kind of molecular computer. Of course there's much more to life than that, but DNA is central to any discussion of life on Earth. DNA is life's most important molecular-level 'space elevator' – a platform from which life can launch itself into higher realms.

The complexity of living creatures arises not because they are made from some special kind of matter – the now-discredited 'vitalist' theory – but because their matter is organized in an exceedingly intricate fashion. DNA does a lot of the routine 'bookkeeping' that *keeps* living creatures organized. Every cell of (nearly) every living organism contains its genome – a kind of code message written in DNA, which gives that organism a lot of hints about how to behave at the molecular level. (Exceptions are various viruses, on the boundary between life and non-life, which use a slightly different code.)

This is why it was possible to clone Dolly the Sheep – to take an ordinary cell from an adult sheep and make it grow into another sheep. The trick actually requires *three* adult sheep. First, there's the one from which you take the cell: call her 'Dolly's Mum'. Then you persuade the cell's nucleus to forget that it came from an adult and to think that it's back in the egg, and then you implant it into an egg from a second sheep ('Egg Donor'). Then you put the egg into the uterus of the third sheep ('Surrogate Mum') so that it can grow into a normal lamb.

Dolly is often said to have been a perfect copy of Dolly's Mum, but that's not completely true. For a start, certain parts of Dolly's DNA came not from Dolly's Mum, but from Egg Donor. And even if that slight difference had been fixed, Dolly could still differ in many ways from her 'mother', because sheep DNA is *not* a complete list of instructions for 'how to build a sheep'. DNA is more like a recipe – and it assumes you already know how to set up your kitchen. So the recipe doesn't say 'put the mixture in a greased pan and place in an oven set to 400°F,' for instance: it says 'put the mixture in the oven' and *assumes* that you know it needs to go in a pan and that the oven should be set to a standard temperature. In particular, sheep DNA leaves out the vital instruction 'put the mixture inside a sheep', but that's the only place (as yet) where you can turn a fertilized sheep egg into a lamb. So even Surrogate Mum played a considerable role in determining what happened when the DNA recipe for Dolly was 'obeyed'.

Many biologists think that this is a minor objection – after all, Egg Donor and Surrogate Mum work the way they do because *their* DNA contains the information that makes them do it. But things that aren't in *any* organism's DNA may be essential for the reproductive cycle. A good example occurs in yeast, a plant that can turn sugar into alcohol and give off carbon dioxide. The entire DNA code for one species of yeast is now known. Thousands of experimentalists have played genetic games with yeast, then spun the beasties in a centrifuge to separate the DNA, from which they can work out the code. When you do this, you leave a scummy residue in the bottom of the test tube, but since it's not DNA, you know it can't be important for genetics, and you throw it away. And so they all did, until in 1997 one geneticist asked a stupid question. If it's not DNA, what's it *for*? What's in that scummy residue, anyway?

The answer was simple, and baffling. Prions. Lots and lots of them.

A prion is a smallish protein molecule that can act as a catalyst for the formation of more protein molecules just like itself. Unlike DNA, it doesn't do this by replication. Instead, it needs a supply of proteins that are *almost* like itself, but not quite – the right atoms, in the right order, but folded into the wrong shape. The prion attaches itself to such a protein, jiggles it around a bit, and nudges it into the same shape as the prion. So now you've got *more* prions, and the process speeds up.

Prions are molecular preachers: they make more of themselves by converting the heathen, not by splitting into identical twins. The most notorious prion is the one that is believed to be the cause of BSE, 'mad cow disease'. The protein that gets converted happens to be a key component of the cow's brain, which is why infected cows lose coordination, stagger around, foam at the mouth, and look crazy. What does yeast want prions for? Without prions, yeast can't reproduce. The protein-making instructions in its DNA sometimes make a protein that is folded into the wrong shape. When a yeast cell divides, it copies its DNA to each half, but it shares the prions (which can be topped up by converting other proteins). So here's a case where, even on the molecular level, an organism's DNA does *not* specify everything about that organism.

There's a lot about the DNA code system that we don't understand, but one part that we *do* is the 'genetic code'. Some segments of DNA are recipes for proteins. In fact, they come very close to being exact blueprints for proteins, because they list the precise components of the protein and they list them in exactly the right order. Proteins are made from a catalogue of fairly tiny molecules known as amino acids. For most organisms, humans included, the catalogue contains exactly 22 amino acids. If you string lots of amino acids together in a row, and let them fold up into a relatively compact tangle, you get a protein. The one thing the DNA doesn't list is *how* to fold the resulting molecule up, but usually it folds the right way of its own accord. Occasionally, when it doesn't, there are more servant molecules to nudge it into the right shape. Just such a servant molecule, rejoicing in the name HSP90, is turning molecular genetics upside down even as we write. HSP90 'insists' that proteins fold into the orthodox shape, even if there are a few mutations in the DNA that codes for those proteins. When the organism is 'stressed', diverting HSP90 to other functions, these cryptic mutations suddenly get expressed – the proteins acquire the unorthodox shape that goes along with their mutated DNA codes. In effect, this says that you can trigger a genetic change by non-genetic means.

Segments of DNA that code for working proteins are called genes. Segments that don't rejoice in a variety of names. Some of them code for proteins that control when a given gene 'switches on', that is, starts to make proteins: these are known as regulatory (or homeotic) genes. Some bits are colloquially called 'junk DNA', a scientific term meaning 'we don't know what these bits are for'. Some literally minded scientists read this as 'they're not *for* anything', thereby getting the horse of nature neatly aligned with the rear end of the cart of human understanding. Most likely they are a mix of different things: DNA that used to have some function way back in evolution but currently does not (and might possibly be revived if, say, an ancient parasite reappeared), DNA that controls how genes switch their protein manufacturing on and off, DNA that controls *those*, and so on. Some may actually be genuine junk. And some (so the joke goes) may encode a message like 'It was me, I'm God, I existed all along, ha ha.'

Evolutionary processes do not always direct themselves along paths that are neatly comprehensible to humans. This doesn't mean Darwin was wrong: it means that even when he's right, there may be a surprising absence of narrativium, so that a 'story' that makes perfect sense to evolution may not make sense to humans. We suspect that a lot of what you find in living organisms is like that – offering a small advantage at every stage of its evolution, but an advantage in such a complex game is that we can't tell a convincing story about *why* it's an advantage. To show just how bizarre evolutionary processes can be, even in comparatively simple circumstances, we must look not to animals or plants, but to electronic circuits.

Since 1993 an engineer named Adrian Thompson has been evolving circuits. The basic technique, known as 'genetic algorithms', is quite widely used in computer science. An algorithm is a specific program, or recipe, to solve a given problem. One way to find algorithms for really tough problems is to 'cross-breed' them and apply natural selection. By 'cross breed' we mean 'mix parts of one algorithm with parts of the other'. Biologists call this 'recombination' and each sexual organism – like you – recombines its parents' chromosomes in just this manner. Such a technique, or its result, is called a genetic algorithm. When the method works, it works brilliantly; its main disadvantage is that you can't always give a sensible explanation of how the resulting algorithm accomplishes whatever it does. More of that in a moment: first we must discuss the electronics.

Thompson wondered what would happen if you used the genetic algorithm approach on an electronic circuit. Decide on some task, randomly cross-breed circuits that might or might not solve it, keep the ones that do better than the rest, and repeat for as many generations as it takes.

Most electronic engineers, thinking about such a project, will quickly realize that it's silly to use genuine circuits. Instead, you can simulate the circuits on a computer (since you know exactly how a circuit behaves) and do the whole job more quickly and more cheaply in simulation. Thompson mistrusted this line of argument, though: maybe real circuits 'knew' something that a simulation would miss.

He decided on a task: to distinguish between two input signals of different frequencies, 1 kilohertz and 10 kilohertz – that is, signals that made 1000 vibrations per second and 10,000 vibrations per second. Think of them as sound: a low tone and a high tone. The circuit should accept the tone as input signal, process it in some manner to be determined by its eventual structure, and produce an output signal. For the high tone, the circuit should output a steady zero volts – that is, no output at all – and for the low tone, the circuit should output a steady five volts. (Actually, these properties were not specified at the start: any two different steady signals would have been acceptable. But that's how it ended up.)

It would take forever to build thousands of trial circuits by hand, so he employed a 'field-programmable gate array'. This is a microchip that contains a number of very tiny transistorized 'logic cells' – mildly intelligent switches, so to speak – whose connections can be changed by loading new instructions into the chip's config-uration memory.

Those instructions are analogous to an organism's DNA code, and can be cross-bred. That's what Thompson did. He started with an array of one hundred logic cells, and used a computer to ran-domly generate a population of fifty instruction codes. The computer loaded each set into the array, fed in the two tones, looked at the outputs, and tried to find some feature that might help in evolving a decent circuit. To begin with, that feature was anything that didn't look totally random. The 'fittest' individual in the first generation produced a steady five-volt output no matter which tone it heard. The least fit instruction codes were then killed off (deleted), the fit ones were bred (copied and recombined), and the process was repeated.

What's most interesting about the experiment is not the details, but how the system homed in on a solution – and the remarkable nature of that solution. By the 220th generation, the fittest circuit produced outputs that were pretty much the same as the inputs, two waveforms of different frequencies. The same effect could have been obtained with no circuit at all, just a bare wire! The desired steady output signals were not yet in prospect.

By the 650th generation, the output for the low tone was steady, but the high tone still produced a variable output signal. It took until generation 2800 for the circuit to give approximately steady, and different, signals for the two tones; only by generation 4100 did the odd glitch get ironed out, after which point little further evolu-tion occurred.

The strangest thing about the eventual solution was its struc-ture. No human engineer would ever have invented it. Indeed no human engineer would have been able to find a solution with a mere 100 logic cells. The human engineer's solution, though, would have been comprehensible – we would be able to tell a convincing 'story' about why it worked. For example, it would include a 'clock' – a circuit that ticks at a constant rate. That would give a baseline to compare the other frequencies against. But you can't make a clock with 100 logic cells. The evolutionary solution didn't *bother* with a clock. Instead, it routed the input signal through a complicated series of loops. These presumably generated time-delayed and otherwise processed versions of the signals, which eventually were combined to produce the steady outputs. Presumably. Thompson described how it functioned like this: 'Really, I don't have the faintest idea how it works.'

Amazingly, further study of the final solution showed that only 32 of its 100 logic cells were actually needed. The rest could be removed from the circuit without affecting its behaviour. At first it looked as if five other logic cells could be removed – they were not connected electrically to the rest, nor to the input or output. However, if these were removed, the circuit ceased to work. Presumably these cells reacted to physical properties of the rest of the circuit other than electrical current – magnetic fields, say. Whatever the reason, Thompson's hunch that a real silicon circuit would have more tricks up its sleeve than a computer simulation turned out to be absolutely right.

The technological justification for Thompson's work is the possibility of evolving highly efficient circuits. But the message for basic evolutionary theory is also important. In effect, it tells us that evolution has no need for narrativium. An evolved solution may 'work' without it being at all clear how it does whatever it does. It may not follow any 'design principle' that makes sense to human beings. Instead, it can follow the emergent logic of Ant Country, which can't be captured in a simple story.

Of course, evolution may sometimes hit on 'designed' solutions, as happens for the eye. Sometimes it hits on solutions that do have a narrative, but we fail to appreciate the story. Stick insects look like sticks, and their eggs look like seeds. There is a kind of Discworld logic to this, since seeds are the 'eggs' of sticks, and prior to the theory of evolution taking hold the Victorians approved of this 'logic' because it looked like God being consistent. The early evolutionists didn't see it that way, and they worried about it; but they worried a lot more when they found that some stick insect eggs looked like little snails. It seemed silly for anything to resemble the favourite food of nearly everything else. In fact, it seemed to be a flat contradiction to the evolutionary story. The puzzle was solved only in 1994, after forest fires in Australia. When new plant shoots came up out of the ashes, they were covered in baby stick insects. Ants had carried the 'seeds', and the 'baby snails', down into their subterranean nests, thinking they were the real thing. Being safely underground, the stick insect eggs escaped the fires. In fact, baby stick insects look, and run, just like ants: this should have been a clue, but nobody made the connection.

And sometimes evolution's solution *has* no narrative structure. To test Darwin's theories thoroughly, we should be looking for evolved systems that *don't* conform to a simple narrative description, as well as for ones that do. Many of the brain's sensory systems may well be like this. The first few layers of the visual cortex, for example, perform generalized functions like detecting edges, but we have no idea how lower layers work, and that may well be because they don't conform to any design principles that we currently can recognize. Our sense of smell seems to be 'organized' along very strange lines, not at all as clearly structured as the visual cortex, and it too may be lacking any element of design.

More importantly, genes may well be like this. Biologists habitually talk of 'the function of a gene' – what it does. The unspoken assumption is that it does only one thing, or a small list of things. This is pure magic: the gene as a spell. It is conceived as being a spell in the same sense that 'Cold Start' in a car is. But a lot of genes may not do *anything* that can be summed up in a simple story. The job they evolved to do is 'build an organism', and they evolved as a team, like Thompson's circuits. When evolution turns up solutions of this kind, conventional reductionism is not much help in understanding those solutions. You can list neural connections till the cows come home, but you won't understand how the cows' visual systems distinguish a cowshed from a bull.

TWENTY-SEVEN

ꞶE ꞐEED
MORE BLOBS

Rᴉɴᴄᴇᴡɪɴᴅ ᴡᴀѕ ꜰɪɴᴅɪɴɢ, now that he was back at what appeared to be his real size, that he was coming to enjoy this world after all. It was so marvellously dull.

Every so often he'd be moved forward a few tens of millions of years. The sea levels would change. There seemed to be more land around, speckled with volcanoes. Sand was turning up on the edge of the sea. Yet the sheer vast ringing silence dominated everything. Oh, there'd be storms, and at night there were brilliant meteor showers that practically hissed across the sky, but these only underlined the absent symphony of life.

He was rather pleased with 'symphony of life'.

'Mr Stibbons?' he said.

'Yes?' said Ponder's voice in his helmet.

'There seem to be a lot of comets about.'

'Yes, they seem to go with roundworld systems. Is this a problem?'

'Aren't they going to crash into this world?'

Rincewind heard the muted sounds of debate in the background, and then Ponder said: 'The Archchancellor says snowballs don't hurt.'

'Oh. Good.'

'We're going to move you on a few million years now. Ready?'

'Millions and millions of years of dullness,' said the Senior Wrangler.

'There are more blobs today,' said Ponder.

'Oh, good. We *need* more blobs.'

There was a yell from Rincewind. The wizards rushed to the omniscope.

'Good heavens,' said the Dean. 'Is that a higher lifeform?'

'I *think*,' said Ponder, 'that seat cushions have inherited the world.'

They lay in the warm shallow water. They were dark green. They were reassuringly dull.

But the *other* things weren't.

Blobs drifted over the sea like giant eyeballs, black, purple, and green. The water itself was covered with them. A scum of them rolled in the surf. The aerial ones bobbed only a few inches above the waves, thick as fog, overshadowing one another in their fight for height.

'Have you *ever* seen anything like that?' said the Senior Wranger.

'Not legally,' said the Dean. A blob burst. Audio reception on the omniscope was not good, but the sound was, in short, *phut*. The stricken thing disappeared into the sea, and the floating blobs closed in over it.

'Get Rincewind to try to communicate with them,' said Ridcully.

'What have blobs got to talk about, sir?' said Ponder. 'Besides, they're not making any noise. I don't think *phut* counts.'

'They're various colours,' said the Lecturer in Recent Runes. 'Perhaps they communicate by changing colour? Like those sea creatures –' He snapped his fingers as an aid to memory.

'Lobsters?' the Dean supplied.

'Really?' said the Senior Wrangler. 'I didn't know they did that.'

'Oh yes,' said Ridcully. 'Red means "help!"'*

'No, I think the Lecturer in Recent Runes is referring to squid,' said Ponder, who knew that this sort of thing could go on for a long time. He added hurriedly, 'I'll tell Rincewind to give it a try.'

Rincewind, apparently knee deep in blobs, said: 'What do you mean?'

'Well … could you get embarrassed, perhaps?'

---

* Wizards seldom bothered to look things up if they could reach an answer by bickering at cross-purposes.

'No, but I'm getting angry!'

'That might work, if you get red enough. They'll think you want help.'

'Do you know there's something else here besides blobs?'

Some of the blobs trailed strands in the faint breeze blowing across the beach. When they tangled up on a blob gasbag, which put some stress on the line, the little blob on the end let go its grip on a rock, the line gradually shortened, and the gasbag bobbed onwards with its new passenger.

Rincewind saw them on a number of blobs. The blobs did not look healthy.

'Predators,' Ponder told him.

'I'm on a beach with *predators*?'

'If it really worries you, try not to look blobby. We'll keep an eye on them. Er … the Faculty is of the opinion that intelligence is most likely to arise in creatures that eat lots of things.'

'Why?'

'Probably because *they* eat lots of things. We'll try a few big jumps in time, all right?'

'I suppose so.'

The world flickered …

'Blobs.'

… flickered …

'The sea's a lot further away. There's a few floating blobs. More black blobs this time.'

… flickered …

'Well out at sea, great rafts of purple blobs, some blobs in the air …'

… flickered …

'Great steaming piles of onions!'

'What?' said Ponder.

'I know it! I just know it! This whole damn place was just lulling me into a false sense of security!'

'What's happening?'

'It's a snowball. The whole world's a giant snowball!'

TWENTY-EIGHT

THE ICEBERG COMETH

 THE EARTH HAS BEEN A GIANT SNOWBALL on many occasions. It was a snowball 2.7 billion years ago, 2.2 billion years ago, and 2 billion years ago. It was a really cold snowball 700 million years ago, and this was followed by a series of global cold snaps that lasted until 600 million years ago. It reverted to snowball mode 300 million years ago, and has been that way on and off for most of the last 50 million years. Ice has played a significant part in the story of life. Just *how* significant a part, we are now beginning to appreciate.

We first began to realize this when we found evidence of the most recent snowball. About one and a half million years ago, round about the time that humans began to become the dominant species on Earth, the planet got very cold. The old name for this period was the Ice Age. We don't call it that any more because it wasn't *one* Age: we talk of 'glacial-interglacial cycles'. Is there a connection? Did the cold climate drive the naked ape to evolve enough intelligence to kill other animals and use their fur to keep warm? To discover and use fire?

This used to be a popular theory. It's possible. Probably not, though: there are too many holes in the logic. But a much earlier, and much more severe, Ice Age very nearly put a stop to the whole of that 'life' nonsense. And, ironically, its failure to do so may have unleashed the full diversity of life as we now know it.

Thanks to the pioneering insights of Louis Agassiz, Victorian scientists knew that the Earth had once been a lot colder than it is now, because they could see the evidence all around them, in the form of the shapes of valleys. In many parts of the world today you can find glaciers – huge 'rivers' of ice, which flow, very slowly, under the pressure of new ice forming further uphill. Glaciers carry large quantities of rock, and they gouge and grind their way along, forming valleys whose cross-section is shaped like a smooth U. All over Europe, indeed over much of the world, there are identical valleys – but no sign of ice for hundreds or thousands of miles. The Victorian geologists pieced together a picture that was a bit worrying in some ways, but reassuring overall. About 1.6 million years back, at the start of the Pleistocene era, the Earth suddenly became colder. The ice caps at the poles advanced, thanks to a rapid build-up of snow, and gouged out those U-shaped valleys. Then the ice retreated again. Four times in all, it was thought, the ice had advanced and retreated, with much of Europe being buried under a layer of ice several miles thick.

Still, there was no need to worry, the geologists said. We seemed to be safe and snug in the middle of a warm period, with no prospect of being buried under miles of ice for quite some time ...

The picture is no longer so comfortable. Indeed, some people think that the greatest threat to humanity is not global warming, but an incipient ice age. How ironic, and how undeserved, if our pollution of the planet cancels out a natural disaster!

As usual, the main reason we now know a lot more is that new kinds of observation became possible, propped up by new theories to explain what it is that they measure and why we can be reasonably sure that they do. These new methods range from clever methods for dating old rocks to studies of the proportions of different isotopes in cores drilled from ancient ice, backed up by ocean-drilling to study the layers of sediment deposited on the sea floor. Warm seas sustain different living creatures, whose death deposits different sediment, so there is a link from sediments to climate.

All of these methods reinforce each other, and lead to very much the same picture. Every so often the surface of the Earth begins to cool, becoming 10°–15°C colder near the poles and 5°C colder elsewhere. Then it suddenly warms up, possibly becoming 3°C warmer than the current norm. In between big fluctuations, there are smaller ones: 'mini ice ages'. The typical gap between a decent-sized ice age and the next is around 75,000 years, often less – nothing like the comfortable 400,000 years of 'interglacial' expected by the Victorians. The most worrying finding of all is that periods of high temperatures – that is, like we get now – seldom lasted more than 20,000 years.

The last major glaciation ended 18,000 years ago.

Wrap up well, folks.

What caused the ice ages? It turns out that the Earth isn't quite as nice a planet as we like to think, and its orbit round the sun isn't quite as stable and repetitive as we usually assume. The currently accepted theory was devised in 1920 by a Serbian called Milutin Milankovitch. In broad terms, the Earth goes round the sun in an ellipse, almost a circle, but there are three features of the Earth's motion that change. One is the amount through which the Earth's axis tilts – about 23° at the moment, but varying slightly in a cycle that lasts roughly 41,000 years. Another is a change in the position of Earth's closest approach to the sun, which varies in a 20,000-year cycle. The third is a variation in the eccentricity of the Earth's orbit – how oval it is – whose period is around 100,000 years. Putting all three cycles together, it is possible to calculate the changes in heat received from the sun. These calculations agree with the known variations in the Earth's temperature, and it seems particularly likely that the Earth's warming up after ice ages is due to increased warmth from the sun, thanks to these three astronomical cycles.

It may seem unsurprising that when the Earth receives more heat from the sun, it warms up, and when it doesn't, it cools down, but not all of the heat that reaches the upper atmosphere gets down to the ground. It can be reflected by clouds, and even if it gets to ground level it can be reflected from the oceans and from snow and ice. It is thought that during ice ages, this reflection causes the Earth to lose more heat than it would otherwise do, so ice ages automatically make themselves *worse*. We get kicked out of them when the incoming heat from the sun is so great that the ice starts to melt despite the lost heat. Or maybe the ice gets dirty, or … It's not so clear that we get kicked into an ice age when less of the sun's warmth reaches the Earth – indeed the slide into an ice age is usually more gradual than the climb back out of it.

All of which makes one wonder whether global warming caused by gases excreted from animals might be partly responsible. When gases such as carbon dioxide and methane build up in the atmosphere, they cause the famous 'greenhouse effect', trapping more sunlight than usual, hence more heat. Right now, most scientists have become convinced, the Earth's supply of 'greenhouse gases' is growing faster than it would otherwise do thanks to human activities such as farming (burning rainforests to clear land), driving cars, burning coal and oil for electricity, and farming again (cows produce a lot of methane: grass goes in one end and methane emerges at the other). And how could we forget the carbon dioxide breathed out by people? One person is equivalent to half a car, maybe more.

Maybe in the past there were vast civilizations of which we now know nothing, all traces having vanished – except for their effect on the global temperature. Maybe the Earth seethed with vast herds of cattle, buffalo, elephants busily excreting methane. But most scientists think that climate change results from variations in five different factors: the sun's output of radiant heat, the Earth's orbit, the composition of the atmosphere, the amount of dust produced by volcanoes, and levels of land and oceans resulting from movement of the Earth's crust. We can't yet put together a really coherent picture in which the measurements match the theory as closely as we'd like, but one thing that *is* becoming clear is that the Earth's climate has more than one 'equilibrium' state. It stays in or near one such state for a while, then switches comparatively rapidly to another, and so on.

The original idea was that one state was a warm climate, like the one we have now, and the other was a cold 'ice age' one. In 1998 Didier Paillard refined this idea to a three-state model: interglacial (warm), mild glacial (coldish), and glacial (very cold). A drop in heat received from the sun below some critical threshold, caused by those astronomical cycles, triggers a switch from warm to coldish. When the resulting ice builds up sufficiently, it reflects so much of the sun's heat that this triggers another switch from coldish to very cold. But when the sun's heat finally builds up again to another threshold value, thanks once more to the three astronomical cycles, then the climate switches back to warm. This model fits observations deduced from the amount of oxygen-18 (a radioactive isotope of oxygen) in geological deposits.

Finally, some drama. About 700 million years ago there was an ice age so severe that it very nearly killed off all of the surface life on Earth. This 'big freeze' lasted for between 10 and 20 million years, the ice reached the equator, and it seems that the seas froze to a depth of half a mile (1 km) or more. According to the 'Snowball Earth' theory, proposed by Paul Hoffman and Daniel Schrag in 1998, ice covered the entire Earth at this time. However, if ice really covered the *whole* Earth, it should have done more damage than the fossil record indicates.

And that's not the only problem. One key piece of evidence for the Big Freeze is a layer of sedimentary rock that formed just after the glaciers melted and left huge quantities of debris. This layer contains less carbon-13, in proportion to ordinary carbon-12, than normal. Marine photosynthesis converts carbon-12 into carbon dioxide more readily than it does carbon-13, so an excess of carbon-13 is left behind in seawater and in the layers of sediment in it that later turn to rock. So a low ratio of carbon-13 to carbon-12 indicates low biological activity.

The scientist's task is to find ways to try to *disprove* things that seem to make sense. In 2001, Martin Kennedy and Nicholas Christie-Blick measured this ratio for sediments that formed *during* the alleged Big Freeze. If the world was miles deep in ice, the ratio ought to be low. But in fact it was high – in Africa, Australia, and North America. This suggests that the global ecosystem was going strong at that time.

Computer models of the climate system show that the oceans strongly resist freezing over completely, too.

Like many attractive scientific theories, Snowball Earth is not at all clear-cut, and further research will be needed to find out who is right. Maybe Earth wasn't a really solid snowball after all. Or maybe, as Schrag responded, there were patches of open water big enough to change the carbon chemistry of the ocean as it absorbed atmospheric carbon dioxide. Maybe the Earth's axis tilted a lot more than astronomers are willing to concede, and the poles lost their ice while equatorial regions gained it. Or perhaps continental drift was more rapid at that time than we think, and we've mapped out the extent of the ice incorrectly. Whatever the details, though, it was a spectacularly icy world.

Although the big freeze came close to wiping out all surface life, it may indirectly have created a lot of today's biodiversity. The big shift from single-celled creatures to multi-celled ones also happened 800 million years ago. It is plausible that the big freeze cleared away a lot of the single-celled lifeforms and opened up new possibilities for multi-celled life, culminating in the Cambrian Explosion 540 million years ago. Mass extinctions are typically succeeded by sudden bursts of diversity, in which life reverts from being a 'professional' at the evolutionary game to being an 'amateur'. It then takes a while for the less able amateurs to be eliminated – and until they are, all sorts of strange strategies for making a living can temporarily thrive. The succession of icy periods that followed the big freeze could only have assisted this process.

However, it may have been the other way round. The invention of the anus by triploblasts may have changed the ecology of the seas. Faeces would have dropped to the sea-bed, where bacteria could specialize in breaking them down. Other organisms could then become filter feeders, living on those bacteria, perhaps sending their larvae up into the plankton for dispersal, as modern filter-feeders do. Several new ways of life depended on this primeval composting system. And it's possible that the successful return of phosphorus and nitrogen into the marine cycles led to an explosion of algae, which reduced atmospheric carbon dioxide, cut back on the greenhouse effect, and triggered the big freeze.

Fortunately for us, the big freeze wasn't *quite* long enough, or cold enough, to kill off everything. (Bacteria in volcanic vents on the ocean floor and in the Earth's crust would have survived no matter what, but evolution would have been set back a long, long way.) So when the Earth warmed, life exploded into a fresh, competition-free world. Paradoxically, a major reason why we are here today may be that we very nearly weren't. Our entire evolutionary history is full of these good news–bad news scenarios, where life leaps forward joyously over the bodies of the fallen …

Rincewind can be forgiven for feeling that Roundworld has it in for him. Life has suffered from many different kinds of natural disaster. Here are two more. In the Permian/Triassic extinction of 250 million years ago, 96% of all species died within the space of a few hundred thousand years.* William Hobster and Mordeckai Magaritz think this happened because they suffocated. Carbon isotopes show that a lot of coal and shale oxidized in the run-up to the extinction, probably because of a fall in sea level, which exposed more land. The result was a lot more carbon dioxide and a lot less oxygen, which was reduced to half today's level. Land species were especially badly affected.

Another global extinction, though less severe, occurred 55 million years ago: the Palaeocene/Eocene boundary. In cores of sediment drilled from the Antarctic, James Kennett and Lowell Stott discovered evidence of the sudden death of a lot of marine species. It seemed that trillions of tons (tonnes) of methane had burst from the ocean, sending temperatures through the roof, methane being a powerful greenhouse gas. Jenny Dickens suggested that the methane was released from deposits of methane hydrates in permafrost and on the seabed. Methane hydrates are a crystal lattice of water enclosing methane gas: they are created when bacteria in mud release the gas and it becomes trapped.

Coincidentally, one of the main results of the Palaeocene/Eocene extinction was a burst of evolutionary diversity, leading in particular to the higher primates – and us. Whether something is a disaster depends on your point of view. Rocks may not have a point of view, as Ponder Stibbons pointed out, but we certainly do.

---

* To the best of our knowledge, based on deduction from the available evidence. Certainly it was a *big* extinction – far bigger than the one that killed off (or helped to kill off) the dinosaurs. We remember the dinosaur one because they've had such good PR people.

TWENTY-NINE
GOING FOR
A PADDLE

I THINK IT LOOKS MORE LIKE A HOGSWATCHNIGHT ORNAMENT,' said the Senior Wrangler later, as the wizards took a pre-dinner drink and stared into the omniscope at the glittering white world. 'Quite pretty, really.'

'Bang go the blobs,' said Ponder Stibbons.

'Phut,' said the Dean, cheerfully. 'More sherry, Archchancellor?'

'Perhaps some instability in the sun ...' Ponder mused.

'Made by unskilled labour,' said Archchancellor Ridcully. 'Bound to happen sooner or later. And then it's nothing but frozen death, the tea-time of the gods and an eternity of cold.'

'Sniffleheim,' said the Dean, who'd got to the sherry ahead of everyone else.

'According to HEX, the air of the planet has changed,' said Ponder.

'A bit academic now, isn't it?' said the Senior Wrangler.

'Ah, I've got an idea!' said the Dean, beaming. 'We can get HEX to reverse the thaumic flow in the cthonic matrix of the optimized bi-direction octagonate, can't we?'

'Well, that's the opinion of four glasses of sherry,' said the Archchancellor briskly, to break the ensuing silence. 'However, if I may express a preference, something that isn't complete gibberish would be more welcome next time, please. So, Mister Stibbons, is this the end of the world?'

'And if it is,' said the Senior Wrangler, 'are we going to have a lot of heroes turning up?'

'What are you talking about, man?' said Ridcully.

'Well, the Dean seems to think we're like gods, and a great many mythologies suggest that when heroes die they go to feast forever in the halls of the gods,' said the Senior Wrangler. 'I just need to know if I should alert the kitchens, that's all.'

'They're only blobs,' said Ridcully. 'What can they do that's heroic?'

'I don't know ... stealing something from the gods is a very classical way,' the Senior Wrangler mused.

'Are you saying we should check our pockets?' said the Archchancellor.

'Well, I haven't seen my penknife lately,' said the Senior Wranger. 'It was just a thought, anyway.'

Ridcully slapped the despondent Stibbons on the back.

'Chin up, lad!' he roared. 'It was a wonderful effort! Admittedly the outcome was a lot of blobs with the intelligence of pea soup, but you shouldn't let utter hopeless failure get you down.'

'*We* never do,' said the Dean.

It was after breakfast next day when Ponder Stibbons wandered into the High Energy Magic building. A scene of desolation met his eye. There were cups and plates everywhere. Paper littered the floor. Forgotten cigarettes had etched their charred trails on the edge of desks. A half-eaten sardine, cheese and blackcurrant pizza, untouched for days, was inching its way to safety.

Sighing, he picked up a broom, and went over the tray containing HEX's overnight write-out.

It seemed a lot fuller than he would have expected.

'Not just blobs – there's *all sorts* of stuff! Some of it's *wiggling* ...'

'Is that a plant or is it an animal?'

'I'm sure it's a plant.'

'Isn't it ... walking ... rather fast?'

'I don't know. I've never seen a plant walking before.'

The wizardery of UU was filtering back in the building as the news got around. The senior members of the faculty were clustered around the omniscope, explaining to one another, now that the impossible had happened, that of course it had been inevitable.

'All those cracks under the sea,' said the Dean. 'And the volcanoes, of course. Heat's bound to build up over time.'

'That doesn't explain all the different shapes, though,' said the Senior Wrangler. 'I mean, the whole sea looks like somebody had just turned over a very big stone.'

'I suppose the blobs had time to consider their future when they were under the ice,' said the Dean. 'It suppose you could think of it as a very long winter evening.'

'I vote for lavatories,' said the Lecturer in Recent Runes.

'Well, I'm sure we all do,' said Ridcully. 'But why at this point?'

'I mean that the blobs were … you know … excusing themselves for millions and millions of years, then you're get a lot of, er, manure …' the Lecturer ventured.

'A shitload,' said the Dean.

'Dean! Really!'

'Sorry, Archchancellor.'

'… and we know dunghills absolutely teem with life …' the Lecturer went on.

'They used to think that rubbish heaps actually generated rats,' said Ridcully. 'Of course, that was just a superstition. It's really seagulls. But you saying life is, as it were, advancing by eating dead men's shoes? Or blobs, in this case. Not shoes, of course, because they didn't have any feet. And wouldn't have been bright enough to invent shoes even if they did. And even if they had been, they couldn't have done. Because there was, at that time, nothing from which shoes might be made. But apart from that, the metaphor stands.'

'There still *are* blobs in there,' said the Dean. 'There's just lots of other things, too.'

'Any of it lookin' intelligent?' said Ridcully.

'I'm not certain how we'd spot that at this stage …'

'Simple. Is anything killing something it doesn't intend to eat?'

They stared into the teeming broth.

'Bit hard to define intentions, really,' said the Dean, after a while.

'Well, does anything look as if it is *about* to become intelligent?'

They watched again.

'That thing like two spiders joined together?' said the Senior

Wrangler after a while. 'It looks very thoughtful.'

'I think it looks very dead.'

'Look, I can see how we can settle this whole evolution business once and for all,' said Ridcully, turning away. 'Mister Stibbons, can HEX use the omniscope to see if anything changes into anything else?'

'Over a moderately sized area, I think he probably can, sir.'

'Get it to pay attention to the land,' said the Dean. 'Is there anything happening on the land?'

'There's a certain greenishness, sir. Seaweed with attitude, really.'

'That's where the interesting stuff will happen, mark my words. I don't know what this universe is using for narrativium, but land's where we'll see any intelligent life.'

'How do you define intelligence?' said Ridcully. 'In the long term, I mean.'

'Universities are a good sign,' said the Dean, to general approval.

'You don't think that perhaps fire and the wheel might be more universally indicative?' said Ponder carefully.

'Not if you live in the water,' said the Senior Wrangler. 'The sea's the place here, I'll be bound. On this world practically nothing happens on the land.'

'But in the water everything's eating each other!'

'Then I'll look forward to seeing what happens to the last one served,' said the Senior Wrangler.

'No, when it comes to universities, the land's the place,' said the Dean. 'Paper won't last five minutes under water. Wouldn't you say so, Librarian?'

The Librarian was still staring into the omniscope.

'Ook,' he said.

'What's that he said?' said Ridcully.

'He said "I think the Senior Wrangler might be right",' said Ponder, going over to the omniscope. 'Oh ... look at *this* ...'

The creature had at least four eyes and ten tentacles. It was using some of the tentacles to manœuvre a slab of rock against another slab.

'It's building a bookcase?' said Ridcully.

'Or possibly a crude rock shelter,' said Ponder Stibbons.

'There we are, then,' said the Senior Wrangler. 'Personal property. Once something is yours, of course you want to improve it. The first step on the road to progress.'

'I'm not sure it's got actual legs,' said Ponder.

'The first slither, then,' said the Senior Wrangler, as the rock slipped from the creatures tentacles. 'We should help it,' he added firmly. 'After all, it wouldn't be here if it wasn't for us.'

'Hold on, hold on,' said the Lecturer in Recent Runes. 'It's only making a shelter. I mean, the Bower Bird builds intricate nests, doesn't it? And the Clock Cuckoo even builds a clock for its mate, and no one says they're *intelligent* as such.'

'Obviously not,' said the Dean. 'They never get the numerals right, the clocks fall apart after a few months, and they generally lose two hours a day. That doesn't sound like intelligence to *me*.'

'What are you suggesting, Runes?' said Ridcully.

'Why don't we send young Rincewind down again in that virtually-there suit? With a trowel, perhaps, and an illustrated manual on basic construction?'

'Would they be able to see him?'

'Er … gentlemen …' said Ponder, who had been letting the eye of the omniscope drift further into the shallows.

'I don't see why not,' said Ridcully.

'Er … there's a … there's …'

'It's one thing to push planets around over millions of years, but at this level we couldn't even give our builder down there a heavy pat on the back,' said the Dean. 'Even if we knew which part of him was his back.'

'Er … something's puddling, sir! Something's going for a paddle, sir!'

It was probably the strangest cry of warning since the famous 'Should the reactor have gone that colour?' The wizards clustered around the omniscope.

Something had gone for a paddle. It had hundreds of little legs.

THIRTY
UNIVERSALS
AND PAROCHIALS

CHANCE MAY HAVE PLAYED A GREATER ROLE than we imagine in ensuring our presence on the Earth. Not only aren't we the pinnacle of evolution: it's conceivable that we very nearly didn't appear at all. On the other hand, if life had wandered off the particular evolutionary track that led to us, it might well have blundered into something similar instead. Intelligent crabs, for example. Or very brainy net-weaving jellyfish.

We have no idea how many promising species got wiped out by a sudden drought, a collapse of some vital resource, a meteorite strike, or a collision with a comet. All we have is a record of those species that happen to have left fossils. When we look at the fossil record, we start to see a vague pattern, a tendency towards increasing complexity. And many of the most important evolutionary innovations seem to have been associated with major catastrophes ...

When we look at today's organisms, some of them seem very simple while others seem more complex. A cockroach looks a lot simpler than an elephant. So we are liable to think of a cockroach as being 'primitive' and an elephant as 'advanced', or we may talk of 'lower' and 'higher' organisms. We also remember that life has evolved, and that today's complex organisms must have had simpler ancestors, and unless we are very careful we think of today's 'primitive' organisms as being typical of the ancestors of today's complex organisms. We are told that humans evolved from something that looked more like an ape, and we conclude that chimpanzees are more primitive, in an evolutionary sense, than we are.

When we do this, we confuse two different things. One is a kind of catalogue-by-complexity of *today's* organisms. The other is a catalogue-by-time of today's organisms, yesterday's ancestors, the day before's ancestors-of-ancestors, and so on. Although today's cockroach may be primitive in the sense that it is simpler than an elephant, it is *not* primitive in the sense of being an ancient ancestral organism. It can't be: it's *today's* cockroach, a dynamic go-ahead cockroach that is ready to face the challenges of the new millennium.

Although ancient fossil cockroaches have the same appearance as modern ones, they operated against a different backgrounds. What you needed to be a viable cockroach in the Cretaceous was probably rather different from what you need to be a viable cockroach today. In particular, the DNA of a Cretaceous cockroach was probably significantly different from the DNA of a modern cockroach. Your genes have to run very fast in order for your body to stand still.

The general picture of evolution that theorists have homed in on resembles a branching tree, with time rising like the sap from the trunk at the bottom, four billion years in the past, to the tips of the topmost twigs, the present. Each bough, branch, or twig represents a species, and all branches point upwards. This 'Tree of Life' picture is faithful to one key feature of evolution – once a branch has split, it doesn't join up again. Species diverge, but they can't merge.*

However, the tree image is misleading in several respects. There is, for instance, no relation between the thickness of a branch and the size of the corresponding population – the thick trunk at the bottom may represent fewer organisms, or less total organic mass,

* There's a silly reason for this, and a sensible one. The silly reason is that species are usually defined to be different if they don't interbreed. If two separate species don't interbreed, it's difficult to put them back together again. The sensible one is that evolution occurs by random mutations – changes to the DNA code – followed by selection. Once a change has occurred, it's unlikely for it to be undone by further random mutations. It's like driving along country roads at random, reaching some particular place, and then *continuing* at random. What you *don't* expect is to reverse your previous path and end up back where you started.

than the twig at the top. (Think about the human twig …) The way branches split may also be misleading: it implies a kind of long-term continuity of species, even when new ones appear, because on a tree the new branches grow gradually out of the old ones. Darwin thought that speciation – the formation of new species – is generally gradual, but he may have been wrong. The theory of 'punctuated equilibrium' of Stephen Jay Gould and Niles Eldredge maintains the contrary: speciation is sudden. In fact there are excellent mathematical reasons for expecting speciation to have elements of both – sometimes sudden, sometimes gradual.

Another problem with the Tree of Life image is that many of its branches are missing – many species go unrepresented in the fossil record. The most misleading feature of all is the way humans get placed right at the top. For psychological reasons we equate height with importance (as in the phrase 'your royal highness'), and we rather like the idea that we're the most important creature on the planet. However, the height of a species in the Tree of Life indicates when it flourished, so every modern organism, be it a cockroach, a bee, a tapeworm, or a cow, is just as exalted as we are.

Gould, in *Wonderful Life*, objected to the 'tree' image for other reasons, and he based his objections on a remarkable series of fossils preserved in a layer of rock known as the Burgess Shale. These fossils, which date from the start of the Cambrian era,* are the remains of soft-bodied creatures living on mud-banks at the base of an algal reef, which became trapped under a mudslide. Very few fossils of soft-bodied creatures exist, because normally only the harder parts survive fossilization. (Some good deposits are now known in China, too.) However, the significance of the Burgess Shale fossils went unrecognized from their discovery by Charles Walcott in 1909, until Harry Whittington took a closer look at them in 1971. The organisms were all squashed flat, and it was virtually impossible to recognize what shape they'd been while alive. Then Simon Conway Morris teased the squished layers apart, and reconstructed

* According to the most recent dating methods, the Cambrian began 543 million years ago. The Burgess shale was deposited about 530–520 million years ago.

the original forms using a computer – and the strange secret of the Burgess Shale was revealed to the world.

Until that point, palaeontologists had classified the Burgess Shale organisms into various conventional types – worms, arthropods, whatever. But now it became clear that most of those assignments were mistaken. We knew, for example, just four conventional types of arthropod: trilobites (now extinct), chelicerates (spiders, scorpions), crustaceans (crabs, shrimp), and uniramians (insects and others). The Burgess Shale contains representatives of all of these – but it also contains *twenty* other radically different types. In that one mudslide, preserved in layers of shale like pressed flowers in the pages of a book, we find more diversity than in the whole of life today.

Musing on this amazing discovery, Gould realized that most branches of the Tree of Life that grew from the Burgess beasts must have 'snapped off' by way of extinction. Long ago, 20 of those 24 arthropod body plans disappeared from the face of the Earth. The Grim Reaper was pruning the Tree of Life, and being heavy-handed with the shears. So Gould suggested that a better image than a tree would be something like scrubland. Here and there 'bushes' of species sprouted from the primal ground level. Most, however, ceased to grow, and were pruned to a standstill hundreds of millions of years ago. Other bushes grew to tall shrubs before stopping … and one tall tree made it right up to the present day. Or maybe we've reconstructed it incorrectly, amalgamating several different trees into one.

This new image changes our view of human evolution. One animal in the Burgess Shale, named *Pikaia*, is a chordate. This is the group that evolved into all of today's animals that have a spinal cord, including fishes, amphibians, reptiles, birds, and mammals. *Pikaia* is our distant ancestor. Another creature in the Burgess Shale, *Nectocaris*, has an arthropod-like front end but a chordate back, and it has left no surviving progeny. Yet they both shared the same environment, and neither is more obviously 'fit' to survive than the other. Indeed, if one had been less evolutionarily fit, it would almost certainly have died out long before the fossils were formed. So what determined which branch survived and which didn't? Gould's suggestion was: *chance*.

The Burgess Shale formed on a major geological boundary: at the end of the Precambrian era and the start of the Palaeozoic. The early part of the Palaeozoic is known as the Cambrian period, and it is a time of enormous biological diversity – the 'Cambrian explosion'. The Earth's creatures were recovering from the mass extinction of the Ediacarans, and evolution took the opportunity to play new games, because for a while it didn't matter much if it played them badly. The 'selection pressure' on new body-plans was small because life hadn't fully recovered from the big die-back. In these circumstances, said Gould, what survives and what does not is mostly a matter of luck – mudslide or no mudslide, dry climate or wet. If you were to re-run evolution past this point, it's quite likely that totally different organisms would survive, different branches of the Tree of Life would be snipped off.

Second time round, it could easily be *our* branch that got pruned.

This vision of evolution as a 'contingent' process, one with a lot of random chance involved, has a certain appeal. It is a very strong way to make the point that humans are *not* the pinnacle of creation, *not* the purpose of the whole enterprise.* How could we be, if a few random glitches could have swept us from the board altogether? However, Gould rather overplayed his hand (and he backed off a bit in subsequent writings). One minor problem is that more recent reconstructions of the Burgess Shale beasts suggests that their diversity may have been somewhat overrated – though they were still very diverse.

But the main hole in the argument is convergence. Evolution settles on solutions to problems of survival, and often the range of solutions is small. Our present world is littered with examples of 'convergent evolution', in which creatures have very similar forms

---

* In the words of Discworld's God of Evolution: 'The purpose of the whole thing is to *be* the whole thing.'

but very different evolutionary histories. The shark and the dolphin, for instance, have the same streamlined shape, pointed snout, and triangular dorsal fin. But the shark is a fish and the dolphin is a mammal.

We can divide features of organisms into two broad classes: universals and parochials. Universals are general solutions to survival problems – methods that are widely applicable and which evolved independently on several occasions. Wings, for instance, are universals for flight: they evolved separately in insects, birds, bats, even flying fish. Parochials happen by accident, and there's no reason for them to be repeated. Our foodway crosses our airway, leading to lots of coughs and splutters when 'something goes down the wrong way'. This isn't a universal: we have it because it so happened that our distant ancestor who first crawled out of the ocean had it. It's not even a terribly sensible arrangement – it just works well enough for its flaws not to count against us when combined with everything else that makes us human. Its deficiencies were tolerated from the first fish-out-of-water, through amphibians and dinosaurs, to modern birds, and from amphibians through mammal-like reptiles to mammals like us. Because evolution can't easily 'un-evolve' major features of body-plans, we're stuck with it.

If our distant ancestors had got themselves killed off by accident, would anything like us still be around? It seems very unlikely that creatures exactly like us would have turned up, because a lot of what makes us tick is parochials. But intelligence looks like a clear case of a universal – cephalopods evolved intelligence independently of mammals, and anyway, intelligence is such a generic trick. So it seems likely that some other form of intelligent life would have evolved instead, though not necessarily adhering to the same timetable. On an alternative Earth, intelligent crabs might invent a fantasy world shaped like a shallow bowl that rides on six sponges on the back of a giant sea urchin. Three of them could at this very moment be writing *The Science of Dishworld*.

Sorry. But it is true. But for a fall of rock here, a tidal pattern there, we wouldn't have been us. The interesting thing is that we almost certainly *would* have been something else.

THIRTY-ONE

GREAT LEAP SIDEWAYS

 RINCEWIND WAS IN HIS NEW OFFICE, filing rocks. He'd worked out quite a good system, based on size, shape, colour and twenty-seven other qualities including whether or not he felt that it was a friendly sort of rock.

With careful attention to cross-referencing, he reckoned that dealing with just those rocks in this room would take him at least three quiet, blessed years.

And he was therefore surprised to find himself picked up bodily and virtually carried towards the High Energy Magic building holding, in one hand, a hard square light grey rock and, in the other hand, a rock that appeared to be well disposed to people.

'Is this *yours*?' roared Ridcully, stepping side to reveal the omniscope.

The Luggage was now bobbing contently a few metres offshore.

'Er ...' said Rincewind. 'Sort of mine.'

'So how did it get *there*?'

'Er ... it's probably looking for me,' said Rincewind. 'Sometimes it loses track.'

'But that's another universe!' said the Dean.

'Sorry.'

'Can you call it back?'

'Good heavens, no. If I could call it back, I'd send it away.'

'Sapient pearwood *is* meta-magical and will track its owner absolutely *anywhere* in time and space,' said Ponder.

'Yes, but not this bit!' said Ridcully.

'I don't recall "not this bit" ever being recorded as a valid sub-set of "time and space", sir,' said Ponder. 'In fact, "not this bit" has never even been accepted as a valid part of any magical invocation, ever since the late Funnit the Foregetful tried to use it as a last-minute addition to his famously successful spell to destroy the entire tree he was sitting in.'

'The Luggage may consist of a subset of at least n dimensions which may co-exist with any other set of $>n$ dimensions,' said the Bursar.

'Don't pay any attention, Stibbons,' said Ridcully wearily. 'He's been spouting this stuff ever since he tried to understand HEX's write-out. It's completely gibberish. What's "n", then, old chap?'

'Umpt,' said the Bursar.

'Ah, imaginary numbers again,' said the Dean. 'That's the one he says should come between three and four.'

'There isn't a number between three and four,' said Ridcully.

'He imagines there is,' said the Dean.

'Could we get inside the Luggage in order to physically go into the project universe?' said Ponder.

'You could try,' said Rincewind. 'I personally would rather saw my own nose off.'

'Ah. Really?'

'But the thought occurs,' said Ridcully, 'that we can use it to bring things back. Eh?'

Down under the warm water, the strange creature's stone structure collapsed for the umpteenth time.

A week went past. On Tuesday a left-over snowball collided with the planet, causing considerable vexation to the wizards and destroying an entire species of net-weaving jellyfish of which the Senior Wrangler had professed great hopes. But at least the Luggage *could* be used to bring back any specimens stupid enough to swim into something sitting underwater with its lid open, and this included practically everything in the sea at the moment.

Life in the round world seemed to possess a quality so prevalent that the wizards even discussed the idea that it was some conceptual element, which was perhaps trying to fill the gap left by the non-existent deitygen.

'However,' Ridcully announced, 'Bloodimindium is not a good name.'

'Perhaps if we change the accent slightly,' said the Lecturer in Recent Runes. 'Blod-di-*min*-dium, do you think?'

'They've certainly got a lot of it, whatever we call it,' said the Dean. 'It's not a world to let a complete catastrophe get it down.'

Things turned up. Shellfish suddenly seemed very popular. A theory gaining ground was that the world itself was generating them in some sort of automatic way.

'Obviously, if you have too many rabbits, you need to invent foxes,' said the Dean, at one of the regular meetings. 'If you've got fish, and you want phosphates, you need seabirds.'

'That only works if you have narrativium,' said Ponder. 'We've got no evidence, sir, that anything on the planet has any concept of causality. Things just live and die.'

And then, on Thursday, the Senior Wrangler spotted a fish. A real, swimming fish.

'There you are,' he said triumphantly. 'The seas are the natural home of life. Look at the land. It's just rubbish, quite frankly.'

'But the sea's not *getting* anywhere,' said Ridcully. 'Look at those tentacled shellfish you were trying to educate yesterday. Even if you so much as made a sudden movement they just squirted ink at you and swam away.'

'No, no, they were trying to *communicate*,' the Senior Wrangler insisted. 'Ink is a natural medium, after all. Don't you get the impression that everything is *striving*? Look at them. You can *see* them thinking, can't you?'

There were a couple of the things in a tank behind him, peering out of their big spiral shells. The Senior Wrangler had the idea that they could be taught simple tasks, which they would then pass on to the other ammonites. They were turning out to be rather a disappointment. They might be good at thinking, ran the general view, but they were pants at actually doing anything about it.

'That's because here's no point in being able to think if you haven't got much to think about,' said the Dean. 'Damn all to think about in the sea. Tide comes in, tide goes out, everything's damp, end of philosophical discourse.'

'Now *these* are the chaps,' he went on, strolling along to another tank. The Luggage had been quite good as a collector, provided the specimens didn't appear to be threatening Rincewind.

'Hmph,' sniffed the Senior Wrangler. 'Underwater woodlice.'

'But there's a lot of them,' said the Dean. 'And they have legs. I've seen them on the seashore.'

'By accident. And they haven't got anything to use as hands.'

'Ah, well, I'm glad you've pointed that out ...' said the Dean, walking along to the next aquarium.

It contained crabs.

The Senior Wrangler had to admit that crabs looked a good contender for Highest Lifeform status. HEX had located some on the other side of the world that were moving along very well indeed, with small underwater cities guarded by carefully transplanted sea-anemones and what appeared to be shellfish farms. They had even invented a primitive form of warfare and had built statues, of sand and spit, apparently to famous crabs who had fallen in the struggle.

The wizards went and had another look fifty thousand years later, after coffee. To the Dean's glee, population pressure had forced the crabs on to the land as well. The architecture hadn't improved, but there were now seaweed farms in the lagoons, and some apparently more stupid crabs had been enslaved for transport purposes and use in inter-clan campaigns. Several large rafts with crudely woven sails were moored in one lagoon, and swarming with crabs. It seemed that crabkind was planning a Great Leap Sideways.

'Not *quite* there yet,' said Ridcully. 'But definitely very promising, Dean.'

'You see, water's too *easy*,' said the Dean. 'Your food floats by, there's not much in the way of weather, there's nothing to kick against ... mark my words, the land is the place for building a bit of backbone ...'

There was a clatter from HEX, and the field of vision of the omniscope was pulled back rapidly until the world was just a mur ble floating in space.

'Oh dear,' said the Archchancellor, pointing to a trail of gas, 'Incoming.'

The wizards watched gloomily as a large part of one hemisphere became a cauldron of steam and fire.

'Is this going to happen *every* time?' said the Dean, as the smoke died away and spread out across the seas.

'I blame the over-large sun and all those planets,' said Ridcully. 'And you fellows should have cleared out the snowballs. Sooner or later, they fall in.'

'It'd just be nice for a species to make a go of things for five minutes without being frozen solid or broiled,' said the Senior Wrangler.

'That's life,' said Ridcully.

'But not for long,' said the Senior Wrangler.

There was a whimper from behind them.

Rincewind hung in the air, the outline of the virtually-there suit shimmering around him.

'What's up with him?' said Ridcully.

'Er ... I asked him to investigate the crab civilization, sir.'

'The one the comet just landed on?'

'Yes, sir. A billion tons of rock have just evaporated around him, sir.'

'It couldn't have *hurt* him, though, could it?'

'Probably made him jump, sir.'

THIRTY-TWO

DON'T LOOK UP

 THE WIZARDS HAVE BEEN CONVINCED all along that a planet is not really a good place to put living creatures. A nice, flat disc, with an attendant turtle who can deal with any inbound rocks before they can wreak havoc, makes much more sense.

It looks increasingly as if they're right. The more we learn about the history of our planet, and the greater universe in which it resides, the more we have to admit that the wizards have a point. Not about the shape of our world, but about how dangerously exposed it is without a turtle. The universe is riddled with flying rocks and awash with radiation; most of it is either close to absolute zero or hot enough to make a hydrogen bomb seem like a nice, comfortable bonfire. Yet, somehow, life managed to gain a hold on at least one planet, and to retain that hold for four billion years – despite everything that the universe threw at it. (Often literally.) And despite every kind of nasty that the planet itself managed to concoct.

There are two ways to interpret this.

One is that life is incredibly fragile, and that Earth is one of the few places where the conditions *necessary* for life managed to hold together long enough for life to develop, diversify, and thrive. At any moment some disaster could undo all that good work and wipe the face of the planet clean of living creatures. The crab civilisation is fictitious, of course, but it's in our story to make two important points. First, that there had been plenty of time for lifeforms at least as intelligent as us to evolve on Earth; second, that if they had done, they could easily have left no trace of their existence. Oh, and third... that there are plenty of ways in which they could have come to a sticky end. So we've been incredibly lucky to avoid going the way of the crab civilisation. On millions of other apparently suitable worlds, life was not so lucky; it either never got started, or something wiped it out. Life is a rarity; Earth may be the only place in the entire universe where that fragile miracle happened.

The other is that life is incredibly robust, and that the conditions on Earth are *sufficient* for life to arise, but by no means necessary. Just because things worked out in a particular way here, it would be a mistake to conclude that the same events must happen everywhere else. An important implication of evolution is that life automatically adapts itself to whichever environment happens to be available. Boiling water at the bottom of the ocean? Just what extremophile bacteria need. Two miles down in the rocks? Super – it's nice and warm down there, and there's plenty sulphur and iron to provide energy. Thank providence there's none of that poisonous oxygen; terrible stuff, violently reactive, immensely destructive. *Nothing* could survive in an oxygen atmosphere ...

Both points of view have their advocates, and both have a certain amount going for them. Until we get to other worlds and find out what's there, there will be plenty of room for disagreement and debate. And, perhaps, a synthesis. Already both viewpoints agree that however life arose here, Earth was no Garden of Eden. Our planet is by no means the ideal habitat for life. In order for living creatures to survive, evolution has had to solve a lot of difficult problems, and adapt to hostile conditions.

You may not realise just how hostile. But think of the common disasters: fires, hurricanes, tornadoes, earthquakes, volcanoes, tidal waves, floods, droughts ... too much rain and we're up to our necks in water; too little and our crops won't grow and we starve.

But those are feeble compared to the *big* disasters.

We tend to think of the history of Life on Earth as the smooth growth of a single great Evolutionary Tree. But that image is getting very long in the tooth. The history of life is more like a jungle than a tree, and most of the plants in the jungle were strangled, squashed, or suffocated before they took even the first step on the road to maturity. And however that jungle grew, there was nothing smooth about it.

True, there was a very long time when there were only 'blobs' in the seas, and we might think of that period as a fairly featureless trunk of the Tree. As far as the blobs were concerned, life probably was pretty uneventful – but only because they didn't *notice* what was happening to the planet. They were largely unaffected by a whole succession of events that would have been cosmic catastrophes for later, more complex life.

There were certainly a few pretty big impacts at the beginning of life on the planet that didn't put them out of business, such as it was. And Snowball Earth – if in fact it ever happened – can't have been easy. But despite all these obstacles, or even because of them, life slowly changed, evolved and diversified as the cukaryotes learned to live in an oxygen atmosphere.

That should have been a disaster. The very composition of the atmosphere changed, and all the biochemical tricks that had evolved to suit the available range of gases became obsolete. Worse, the gas polluting the air was oxygen, an appallingly reactive substance. Think of what would happen today if the atmosphere started to be taken over by fluorine. Some of the nastiest, most explosive substances are fluorine compounds. But oxygen is just as bad, if not worse; think of fires, think of rust, think of decay.

The eukaryote cell triumphed over oxygen, and subverted it. Oxygen's negative characteristics were turned into positive ones. So effective was this evolutionary revolution that the deadly, poisonous pollutant became *essential* for (most) life. Deprive a human, a dog, or a fish of its oxygen, and it dies very quickly. Water, food ... those it can do without, for a time. But oxygen? You'll survive for a few minutes at most, maybe half an hour if you're a whale.

The oxygen trick was so good that it took over. Eukaryote life radiated – diversified rapidly – in the seas, inventing entire new kinds of ecologies. With that diversification as a springboard, life came out on land. The advantage of moving to the land was that it opened up an entire range of new habitats, new ways of making a living. So many new kinds of living organism could thrive. One disadvantage, though, was that living on land made life much more vulnerable to astronomical insults. Living on land produced many more complicated kinds of plants and animals, able to protect themselves against small local changes, like hot sunshine, or snow. But, ironically, that very complication made them much more vulnerable to big problems – like stones falling from the sky.

We all know about the meteorite that killed the dinosaurs ... and that fits. Dinosaurs were wonderfully effective as long as the environment remained suitable, but they were not at all well-adapted to the sudden changes that the impact created. But bacteria hardly noticed. If anything, it was a good time for them: they got a lot of extra food for a few hundred years, as the corpses decayed, and then went back to the old boring routine.

We'll say a bit more about the 200 million year reign of the dinosaurs and their friends soon, and indeed about what killed them off. But we need to give you some context first. Simple forms of life can put up with a lot, and did. And they changed the planet, or at least its outer skin, by putting in feedback loops that made it less liable to change.

They started Gaia. This is the name that James Lovelock gave in 1982 to the concept of the Earth as a complex living system – metaphorically, an organism in its own right. The idea has been romanticised into the Earth being a kind of Earth-mother, but what do you expect when you attach the name of a goddess to your new scientific concept? Stripped of the romantic frills, the point is that our planet acts as a single system, and it has evolved mechanisms that keep it functioning effectively. This development is a consequence of innumerable subsystems – organisms, ecologies – evolving mechanisms that keep them functioning effectively. If every member of a team gets better at playing their role within the team, then the team as a whole improves too.

Complexity is a double-edged sword. More complex forms of life find that the ordinary problems of living on a planet are more and more under control ... except for those confounded problems from outside, like meteorites, which can be disastrous.

The Moon, Mercury, Mars, and various satellites are covered in circular craters, some large, some small. Nearly all of those craters, we now know, result from the impact of a big lump of rock or ice or a bit of both. A few are volcanic. Not so long ago most were thought to be caused by volcanoes, but that turned out to be wrong.

Several planets, among them the Earth, do not show obvious signs of impacts. Is that because nothing hit them? No. An atmosphere helps: smaller bodies burn up before they hit the ground. It's the closest to a protector-turtle that we get. But bigger rocks can still get through the defences. The main reason why some planets show no clear signs of impacts is because those planets have weather, like the Earth, which erodes the craters until they disappear, or episodes of massive vulcanism, like Venus, which resurfaces the planet, or are gas to begin with, like Jupiter and Saturn, and don't show permanent marks.

In Quebec there is a lake called Manicouagan. You can't miss it on a map: look near 51°N, 68°W. It's circular, and it's big: 44 miles (71 km) in diameter. It is the weathered remains of a gigantic crater that formed 210 million years ago when a rock two or three miles (3-5km) across collided with the Earth. There is a central peak made from rock that melted in the heat that the impact generated and then solidified; more molten rock flowed across the floor of the crater and still can be found today. The lake fills a ring-shaped valley that glaciers carved out of soft rock that originally formed the crater walls, and was eroded away and collapsed.

Also in Canada is the Sudbury impact structure, the largest on the planet. It is 190 miles (300 km) across and 1.85 billion years old, and the rock that made it was about 20 miles (30km) in diameter, and the energy released in the impact was equivalent to one quadrillion tons of TNT, or about ten million really big hydrogen bombs. In Vredefort, South Africa there is another impact structure of a similar size, formed 2.02 billion years ago. These may not remain the record-holders: an even bigger impact structure, about twice the size, is suspected to exist in the Amirante Basin of the Indian Ocean. Altogether, more than 150 impact structures – remnants of craters – have been found on the Earth's land-masses, and many areas have not yet been thoroughly surveyed. More than half the Earth's surface is ocean, and incoming rocks should hit pretty much at random, so the total number is probably closer to 500.

These are all fairly ancient craters, but there is no good reason to believe that such impacts could not happen again. Big impacts are rarer than small ones, because big lumps of rock are rarer than small ones. Impacts the size of Sudbury or Vredefort should happen about once every billion years. (It should not be a surprise that when such impacts finally did arrive, about two billion years ago, two of them came along together.) Since nothing that size has happened for two billion years, it might seems that we are overdue for another one, but that kind of reasoning is a statistical fallacy. Rare, isolated events usually obey the so-called 'Poisson distribution' of probabilities, and one feature of this distribution is that it 'has no memory'. At any time, whether two major impacts have just happened, or none have happened for ages, the average time to the next one is always the same – in this case, a billion years.

It *could* be a few decades, mind you. But it couldn't be tomorrow, or even next year, because we would have spotted such a body coming by now.

The most recent well known impact on Earth was the Tunguska meteorite, which exploded 4 miles (6km) above Siberia in 1908, causing an explosion that felled trees for more than thirty miles (50 km) around. Craters, or other evidence of even more recent impacts have been found. A double-crater in the Saudi Arabian desert may be only a few centuries old.

Where do all these rocks (and other junk, like ice) come from? Who or what is throwing them at us?

First, some terminology. When you look into the night sky and see a 'shooting star', a glowing streak, that's a meteor. It's not a star, of course: it's a lump of cosmic debris that has hit the Earth's atmosphere at high speed and is burning up because of friction. The debris itself is called a meteoroid, and any part that remains when it has hit is called a meteorite. For convenience, though, we'll generally use the word 'meteorite' for all of these. But we thought we ought to show you that we could have been pedantic if we'd wanted to.

Some of these bodies are mostly rock, and some are mostly ice. And some are a bit of both. Wherever they come from, it's not the Earth. At least, not directly. A few may have been splashed off the Earth by a previous impact and then come back down the next time we ran into them. However they got Up There, that's clearly where they are coming from. What's Up There? The rest of the universe. The closest bit is our own Solar System. So that has to be the most likely culprit. And there's no question that it has a lot of ammunition.

Earlier, we described the Solar System as being rather chaotic: nine planets and a few moons with some quite interesting real estate. We mentioned that there was quite a lot left over after these larger bodies were accounted for. There were relatively small lumps of actual rock in the asteroid belt, but nearly all the 'loose change', after the Sun and planets had been paid for out of the Solar nebula, was lumps of dirty ice.

The biggest collection of these is the Oort Cloud, a vast, very thinly spread mass that lives outside the Solar System 'proper' – that is, further out than Pluto (or Neptune when Pluto gets inside the orbit of Neptune, which can happen). In 1950 Jan Hendrik Oort proposed that the source of most of the comets we see from Earth must be some such cloud, and got it named after him. The main evidence we have for its existence is that comets with very long thin orbits, which are common, must have come from somewhere. The bodies in the Oort cloud range from pebble-sized up to lumps perhaps as big as Pluto.

These comet materials are the usual source of the meteorites we pick up and put into museums, after most of their substance has burnt up in the atmosphere. We're beginning to get an idea of how big the Oort Cloud could be. Its mass is about a tenth that of Jupiter, and it extends way outside Pluto's orbit, perhaps as far as 3 light years – two thirds of the way to the nearest star. This spreads the material through a volume millions of times as great the volume inside Pluto's orbit, our actual planetary Solar System. So the 'cloud' is so rarefied that if you went there, you probably wouldn't see anything.

The gravitational pull of the Sun is tiny at those distances, and the dirty ice lumps barely move along their orbital paths, which are probably close to circles. To the extent that the ice lumps *have* orbits, and don't just drift slowly about, they take millions of years to go once round the Sun. But the universe doesn't let them keep doing that without interference. Oort called his cloud 'A garden, gently raked by stellar perturbations'. As nearer stars, and the pull of the whole galaxy, interact with the Sun's pull, many of these lumps are pulled away from their normal paths.

It turns out that the disturbances need not be as gentle as Oort supposed. About once every 35 million years a star passes *through* the Oort cloud, and havoc ensues. Since the 1970s another source of big disturbances has been recognised: Giant Molecular Clouds. These are huge accumulations of cold hydrogen, where stars and solar systems are born. Their masses can be a million times that of the Sun. They don't have to come anywhere near us to shake ice lumps out of their sedate near-circular orbits in the Oort cloud.

Such disturbances can cause lumps of ice to drift in towards the Solar System. At that point, they become comets. Some probably drift outwards too, but we're not so concerned with those. And comets are the main (but not the only) source of cosmic junk in Earth's backyard.

About a thousand meteoroids bigger than a football hit Earth's atmosphere each day, together with countless millions of smaller ones. And as time passes, we receive some big and some bigger, with the occasional dinosaur-killer. How often do we expect to see such a big one? About once every hundred million years.

There is much more of this kind of junk in the Solar system than we used to think, and it rains down on our planet constantly. Every year, we sweep up about 80,000 tones (tonnes) of it. Nearly all of the debris falling on Earth is little bits, mostly somewhat dried-out icy dirt from the tails of comets. Debris of this kind follows the comet's orbit, marking it out like a gravel path. When the Earth's orbit takes it through this cometary junkheap, some of the gravel burns up in the atmosphere, and we see spectacular light shows: meteor showers. These arrive on particular dates each year as the

Earth passes through that debris. For example the Leonids can be seen in November, and the Perseids in August.

There is a bit of a mystery about the Geminid meteor showers, which come in December, though. They seem to be associated with a (defunct) comet whose perihelion, closest approach to the Sun, is out by Pluto's orbit. And that brings us to another source of impactors: the Kuiper Belt, which is the bit of the Oort Cloud not very far outside Pluto. In fact, Pluto and its satellite Charon are now thought not to be a 'real' planet-and-moon, but only the biggest lump in the Kuiper Belt. These lumps travel in genuine quasi-elliptical orbits, and may be the source of some of the regular comets with shorter orbital periods – like Halley's comet, which returns every 76 years or so.

As well as comets, the asteroids also send rocks our way. Jupiter's gravitational field is strong enough to disturb the asteroids, especially those in certain 'resonant' orbits, with periods that are a simple fraction of Jupiter's – one third or two fifths, say. Of the 8000 or so known asteroids, about one in twenty has an orbit that comes close to that of the Earth, or even crosses it. All of those that cross are potential impactors. Asteroids whose orbits approach the Sun to within a distance of 1.3 times the radius of the Earth's orbit are said to be Earth-approaching asteroids, or *Amors*. The best known of these is Eros. Asteroids whose orbits overlap the Earth's are called *Apollos*. More than 400 Amors and Apollos are known. More worrying are *Atens*, which are Amors too small to be detected easily but still big enough to cause tremendous damage. Most of these probably started out in the main asteroid belt, but were disturbed by Jupiter so that they crossed the orbit of Mars, and were then further disturbed by Mars.

This leaves us with two opposite ways to view Jupiter – perhaps complementary ways. This largest planet has been proposed as the saviour of Earth's life forms on countless occasions, its enormous gravity picking up nearly all of the in-falling rocks and icy lumps – as it did comet Shoemaker-Levy 9 in 1994. But it has also been shown to shake the Asteroid Belt about, possibly causing that dinosaur-killer (if it was actually an asteroid) to hit the Earth.

The message is that a basketball left on a billiard table has a fairly interesting life. Velikovsky, who proposed a wild theory in the fifties that made the Solar System look very much like a snooker table in Biblical times, with Mars moving substantially closer to the Earth and Venus turning from a comet into a planet, wasn't very wrong in principle.

Only in every single detail.

Here's something else to worry about. Out there in the Milky Way galaxy there's a lot of stars. Occasionally one goes nova, rarely supernova, as they explode. There is a sphere of very active radiation leaving such stars. If one went off in our vicinity, up to twenty light years away, say, all higher forms on Earth would be sterilised, at least. The bacteria, especially those deep in the Earth's crust, would survive. They probably wouldn't notice a thing. Wait a few billion years … higher lifeforms could exist in abundance once more.

More worrying still are gamma-ray bursters. Gamma-rays are very short wavelength electromagnetic radiation, such as x-rays. When astronomers managed to develop instruments that could detect such radiation, and put them into satellites, they discovered that two or three times per day the Earth is illuminated by an intense burst of gamma-rays coming from somewhere out in space. These gamma ray bursts seem to be extremely energetic: there is good evidence that the source of one of them was 12 billion light years away. Even a supernova would not be visible from that distance, so gamma-ray bursts have to be caused by something really serious.

What? That's a mystery – perhaps the biggest mystery in today's astronomy. The best bet is a collision between neutron stars. Imagine a binary star – two stars, orbiting their common centre of mass. Suppose they are both neutron stars. As time passes, they lose energy and fall in towards each other. If you wait long enough, they will come so close together that they collide. This, by the way, is likely to be a very messy business, not at all as simple as two tennis balls sticking together and rounding off. They probably break up and reform. So far, all the gamma-ray bursters we've seen are a long, long distance away. But one could light up anywhere. If a pair of neutron stars collapsed on to each other within a hundred light years of Earth, life might survive in the deep seas and the deepest rocks, but the rest of our planet would be dead.

And we wouldn't even see it coming.

Asteroids and comets give you a bit of notice. We have the capability, given a year's run-up time, to tackle small Earth-crossing asteroids now. We can see them coming and plot their arrival. But gamma-rays are electromagnetic: they travel at the speed of light. They could be on their way now: we couldn't know. As soon as we did know, we and our technology would be dead.

Even our own Sun is not trustworthy. The nuclear reactions that make stars burn also make them change, as elements are created or used up, or just reach some critical level that triggers new kinds of reactions. Most stars follow the same series of changes, called the main sequence.

When the Sun first arrives on the main sequence, it is just like our Sun, with a surface temperature of about 6,000 degrees Kelvin, a light output of about 400 septillion watts, and a composition of 73% hydrogen, 25% helium, and 2% everything else. It stays on the main sequence for ten billion years, until nearly all of its hydrogen has been fused into helium. At that point, its core starts to contract, and becomes degenerate – consisting of closely packed neutrons. Outside the core there remains a shell of hydrogen, which continues to undergo nuclear reactions, which cause the outer layers of the star to expand and cool. The star becomes a red giant, between 10 and 100 times as big.

The radius of the Sun now is roughly 450,000 miles (700,000 km). At this stage its surface will probably be somewhere between the orbits of Mercury and Venus, and the Earth will already be in serious trouble. But there is more to come. As the core heats, it ignites a nuclear reaction that turns helium into carbon – the very reaction that allegedly is responsible for the existence of carbon-based lifeforms like us. This 'helium flash' happens very quickly, on astronomical timescales, and it destroys the degeneracy of the core. Now the core can once more sustain nuclear reactions, but now it burns helium. The outer layers of the star shrink, and become hotter.

When the helium in the core is used up, the star again burns its nuclear material in two shells: an inner one that burns helium to make carbon, and an outer one that converts hydrogen to helium. The outer layers expand again, and the star becomes a red giant for the second time. Now the outer layers start to blow away, exposing the hot core. The star loses layer upon layer of its material, and shrinks. Finally, the outer layers are all gone, and the core once more becomes degenerate. The star has become a white dwarf.

Our Sun has about 5.7 billion years left on the main sequence; then: *kerblooie*! Red Giant, and Earth becomes a cinder, or even gets gobbled up completely. But don't lose any sleep over it. The typical lifetime of a species is 5 million years. We'll be long gone.

Planets are not comfortable. Even when life has made its own bed (nice oxygenated atmosphere with an ozone layer to keep out the nasty ultra-violet, nice ooze in the bottom of the oceans, nice long relaxation times for the thermal atmospheric oscillators), there are still plenty of things the Universe can throw at a planet that can still make it a bit fragile. If not kill it altogether.

Which brings us back to our original question. Is life fragile, and have we been extraordinarily lucky? Or is it robust, and therefore common? Is life so adaptable that it can handle virtually anything that the universe sends its way?

Until we can explore other worlds and see what kinds of life, if any, are present, anything we say here has to be speculative. The big difficulty is an 'anthropic principle' point. Suppose that life is incredibly rare, and that on most worlds it never really gets going, or doesn't last long, because of all the disasters lying in wait. Nonetheless, there's a lot of galaxies out there, each having billions or even trillions of stars. Even if the chances of survival are very, very small, occasionally one planet will get lucky. Some proportion of planets *must* get lucky, that's how probability works.

Because life on this world *has* survived, we are therefore one of the lucky ones. It then becomes completely irrelevant how small our chances were. We are not representative. The probability that we survived is certainty, because we *did*. So we cannot reason, from our existence, that the chance of survival has to be fairly large. Whether it is large or small, we are here. So this is a case where the anthropist can legitimately frighten us. Perhaps all planets do get life and, if they're allowed enough time; even extelligent life on a few. But we really could be the only one who's survived to ask the question.

On the other hand ... The very diversity of nasty things that the universe has up its sleeve argues for the adaptability and versatility of life. Life on Earth does not look like a bunch of lucky survivors. It looks like a bunch of tough guys who have overcome every obstacle put in their way. Sure, they took casualties, sometimes severe. But as long as a few survive the battle, pretty soon the planet is covered with life again, because life reproduces – fast. Whatever the disaster, it bounces back.

So far, anyway.

THIRTY-THREE
THE FUTURE
IS NEWT

HEX WAS THINKING HARD AGAIN. Running the little universe was taking much less time than it had expected. It more or less ran itself now, in fact. The gravity operated without much attention, rainclouds formed with no major interference and rained every day. Balls went around one another.

HEX didn't think it was a shame about the crabs going. HEX hadn't thought it was marvellous that the crabs had turned up. HEX thought about the crabs as *something that had happened*. But it had been interesting to eavesdrop on Crabbity – the way the crabs named themselves, thought about the universe (in terms of crabs), had legends of the Great Crab clearly visible in the Moon, passed on in curious marks the thoughts of great crabs, and wrote down poetry about the nobility and frailty of crab life, being totally accurate, as it turned out, on this last point.

HEX wondered: if you have life, then intelligence will arise somewhere. If you have intelligence, then extelligence will arise somewhere. If it doesn't, intelligence hasn't got much to be intelligent *about*. It was the difference between one little oceanic crustacean and an entire wall of chalk.

The machine also wondered if it should pass on these insights to the wizards, especially since they actually lived in one of the world's more interesting outcrops of extelligence. But HEX knew that its creators were infinitely cleverer than it was. And great masters of disguise, obviously.

The Lecturer in Recent Runes had designed a creature.

'Really, all we need is a basic limpet or whelk, to begin with,' he said, as they looked at the blackboard. 'We bring it back here where proper magic works, try a few growth spells, and then let Nature take its course. And, since these extinctions seem to be wiping out everything, it'll gradually become the dominant feature.'

'What's the scale again?' said Ridcully, critically.

'About two miles to the tip of the cone,' said the Lecturer. 'About four miles across the base.'

'Not very mobile, then,' said the Dean.

'The weight of the shell will certainly hamper it, but I imagine it should be able to move its own length in a year, perhaps two.'

'What'll it eat, then?'

'Everything else.'

'Such as ...'

'Everything. I'd advise suction holes around the base here so that it can filter seawater for useful things like plankton.'

'Plankton being – ?'

'Oh, whales, shoals of fish and so on.'

The wizards looked long and hard at the huge cone-shaped object.

'Intelligence?' said Ridcully

'What for?' said the Lecturer in Recent Runes.

'Ah.'

'It will withstand anything except a *direct* hit with a comet, and I estimate it'll have a lifespan of about 500,000 years.'

'And then it'll die?' said Ridcully.

'Yes. I estimate it will, by then, take it twenty-four hours and one second to absorb enough food to last it for twenty-four hours.'

'So after that it will be dead?'

'Yes.'

'Will it know?'

'Probably not.'

'Back to the drawing board, Senior Lecturer.'

Ponder sighed.

'It's no good ducking,' he said. 'That won't help. We're paying special attention to comets. We'll let you know in plenty of time.'

'You've got no idea what it was like!' said Rincewind, creeping along the beach. 'And the noise!'

'Have you seen the Luggage?'

'It certainly made my ears ring, I can tell you!'

'And the Luggage?'

'What? Oh ... gone. Have you *looked* at that side of the planet? There's a whole new set of mountain ranges!'

The wizards had let time run forward for a while after the strike. It made such a depressing mess of everything. Now, drawing on its bottomless reserves of bloodimindium, life was returning in strength. Crabs were already back although, here, at least, they didn't seem inclined to make even simple structures. Perhaps something in their souls told them it'd be a waste of time in the long run.

Rincewind mentally crossed them off the list. Look for signs of intelligence, the Archchancellor had said. As far as Rincewind was concerned, anything *really* intelligent would be keeping out of the way of the wizards. If you saw a wizard looking at you, Rincewind would advise, then you should walk into a tree or say 'dur?'.

All along the beaches, and out below the surf, everything was acting with commendable stupidity.

A soft sound made him look down. He'd almost stepped on a fish.

It was some way from the water line, and squirming across the mud towards a pool of brackish water.

A kind man by nature, Rincewind picked it up gingerly and carried it back to the sea. It flopped around in the shallows for a while and then, to his amazement, inched its way back on to the mud.

He put it back again, in deeper water this time.

Thirty seconds later, it was back on the beach.

Rincewind crouched down, as the thing wiggled determinedly onwards.

'Would it help to talk to someone?' he said. 'I mean, you've got a good life out there in the sea, no sense in throwing it all away, is there? There's always a silver lining if you know where to look. Okay, okay, life's a beach. And you're a pretty ugly fish. But, you know, beauty is only sk– scale deep, and –'

'What's happening?' said Ponder's voice in his ear.

'I was talking to this fish,' said Rincewind.

'Why?'

'It keeps coming out of the water. It seems to want to go for whatever is the opposite of a paddle.'

'Well?'

'You told me to keep a look out for anything interesting.'

'The consensus here is that fish aren't interesting,' said Ponder. 'Fish are dull.'

'I can see bigger fish in the shallows,' said Rincewind. 'Perhaps it's trying to keep away from them?'

'Rincewind, fish are designed for living in water. That's why they're fish. Go and find some crabs. And put the poor freakish thing back in the sea, for goodness' sake.'

'Perhaps a rethink is in order here,' said Ridcully.

'About the newts,' said Ponder.

'Newts is going far too far,' said the Dean. 'I've seen more shapely things in the privy.'

'I want the person who put the newts on this continent to own up right now,' said Ridcully.

'No one could,' said the Senior Wrangler. 'No one's seen the Luggage since the last comet. We couldn't get anything in there.'

'I know, because I had a tank of thaumically treated whelks all ready to go,' said the Lecturer in Recent Runes. 'And what, pray, am I supposed to do with them?'

'Some sort of chowder would appear to be in order,' said the Dean.

'Evolution makes things better,' said Ridcully. 'It can't make them *different*. All right, some rather dull amphibians seem to have turned up. *But*, and this is important, those fish Rincewind reported are still around. Now, if they were going to turn into things with legs, why are they still here?'

'Tadpoles are fish,' said the Bursar.

'But a tadpole *knows* it's going to be a frog,' said Ridcully patiently. 'There's no narrativium on this world. That fish couldn't be saying to itself "Ah, a new life beckons on dry land, walking around on things I haven't yet got a name for."' No, either the planet is somehow generating new life, or we're back to the old "hidden gods" theory.'

'It's all gone wrong, you know,' said the Dean. 'It's the bloodi-mindium. Even gods couldn't control this place. Once there's life, there's complete and utter chaos. Remember that book the Librarian brought back? It's a complete fantasy! Nothing seems to happen like that at all! Everything just does what it likes!'

'Progress is being made,' said Ponder.

'Big amphibians?' sneered the Senior Wrangler. 'And things were going so well in the sea. Remember those jellyfish that made nets? And the crabs even had a flourishing land civilization! They had practically got a culture!'

'They *ate* captured enemies *alive*,' said the Lecturer in Recent Runes, patiently.

'Well ... yes. But with a certain amount of etiquette, at least,' the Senior Wrangler admitted. 'And in front of their sand statue of the Great Big Crab. They were obviously attempting to control their world. And what good did it do them? A million tons of white hot ice smack between the eyestalks. It's so upsetting.'

'Perhaps they should have eaten more enemies,' said the Dean.

'Perhaps sooner or later the planet will get the message,' said Ridcully.

'Time for the giant whelks, perhaps?' said the Lecturer in Recent Runes, hopefully.

'Big newts is what we've got right now,' said Ridcully. He glanced at the Dean and Senior Wrangler. Ridcully hadn't maintained his position atop the boiling heap of UU wizardry without a little political savvy. 'And newts, gentlemen, might be the way to go. Amphibians? At home in the water *and* on land? The best of both worlds, I fancy.'

The two wizards exchanged sheepish glances.

'Well ... I suppose ...' said the Senior Wrangler.

'Could be,' the Dean said grudgingly. 'Could be.'

'There we are then,' said Ridcully happily. 'The future is newt.'

THIRTY-FOUR
NINE TIMES OUT OF TEN

 THERE'S NO NARRATIVIUM ON THIS WORLD.'

Let's take a step away from the unfolding ancestral tale of The Fish That Came Out From The Sea and look at a more philosophical issue. The wizards are puzzled. On Discworld, things happen because narrative imperative *makes* them happen. There is no choice about ends, only about means. The Lecturer in Recent Runes is trying to make a sustainable lifeform happen. He thinks that the obstacle to sustainability is the fragility of life – so the only way he can see to achieve this is the two-mile limpet, proof against everything the sky can drop on it.

It never occurs to him that lifeforms might achieve sustainability by other, less direct methods, despite the evidence of his eyes that suggests that a dogged tenacity appears to allow life to arise in the most inhospitable environment, effectively re-creating itself over and over again. The wizards are torn between evidence that a planet is the last place you'd choose to create life, and evidence that life doesn't agree.

On Discworld, it is clearly recognized that million-to-one chances happen nine times out of ten.* The reason is that every Discworld character lives out a story, and the demands of the story determine how their lives unfold. If a million to one chance is required to keep that story on track, then that's what will happen, appalling odds notwithstanding. On Discworld, abstractions generally show up as *things*, so there is even a thing – narrativium – that ensures that everybody obeys the narrative imperative. Another personification of the abstract, Death, also makes sure that each

---

* Indeed, it is a fundamental part of story telling. If the hero did not overcome huge odds, what would be the *point*?

individual's story comes to an end exactly when it's supposed to. Even if a character tries to behave contrary to the story in which they find themselves, narrativium makes sure that the end result is consistent with the story anyway.

What's puzzling the wizards is that our world isn't like that ... Or is it?

After all, people live on our world too, and it's people that drive stories.

As case in point, a story about people who drive. The setting is Jerez Grand Prix circuit, last race of the 1997–98 Formula One motor racing season ... Ace driver Michael Schumacher is one Championship point ahead of arch-rival Jacques Villeneuve. Villeneuve's team-mate Heinz-Harold Frentzen may well play a crucial tactical role. The drivers are competing for 'pole position' on the starting grid, which goes to whoever produces the fastest lap in the qualifying sessions. So what happens? Unprecedentedly, Villeneuve, Schumacher, and Frentzen all lap in 1 minute 21.072 seconds, the same time to a thousandth of a second. An amazing coincidence.

Well: 'coincidence' it surely was – the lap times *coincided*. But was it truly amazing?

Questions like this arise in science, too, and they're important. How significant is a statistical cluster of leukaemia cases near a nuclear installation? Does a strong correlation between lung cancer and having a smoker in the family really indicate that secondary smoking is dangerous? Are sexually abnormal fish a sign of oestrogen-like chemicals in our water supply?

Another case in point. It is said that 84% of the children of Israeli fighter pilots are girls. What is it about the life of a fighter pilot that produces such a predominance of daughters? Could an answer lead to a breakthrough in choosing the sex of your children? Or is it just a statistical freak? It's not so easy to decide. Gut feelings are worse than useless, because human beings have a rather poor intuition for random events. Many people believe that lottery numbers that have so far been neglected are more likely to come up in future. But the lottery machine has no 'memory' – its future is independent of its past. Those coloured plastic balls *do not know* how often they have come up in previous draws, and they have no tendency to compensate for past imbalances.

Our intuition goes even further astray when it comes to coincidences. You go to the swimming baths, and the guy behind the counter pulls a key at random from a drawer. You arrive in the changing room and are relieved to find that very few lockers are in use ... and then it turns out that three people have been given lockers next to yours, and it's all 'sorry!' and banging locker doors together. Or you are in Hawaii, for the only time in your life ... and you bump into the Hungarian you worked with at Harvard. Or you're on honeymoon camping in a remote part of Ireland ... and you and your new wife meet your Head of Department and *his* new wife, walking the other way along an otherwise deserted beach. All of these have happened to Jack.

Why do we find coincidences so striking? Because we expect random events to be evenly distributed, so statistical clumps surprise us. We think that a 'typical' lottery draw is something like 5, 14, 27, 36, 39, 45, but that 1, 2, 3, 19, 20, 21 is far less likely. Actually, these two sets of numbers have exactly the same probability, which for the UK lottery is: 1 in 13,983,816. A typical lottery draw often includes several numbers close together, because sequences of six random numbers between 1 and 49, which is how the UK lottery works, are more likely to be clumpy than not.

How do we know this? Probability theorists tackle such questions using 'sample spaces' – their name for what we earlier called a 'phase space', a conceptual 'space' that organizes all the possibilities. A sample space contains not just the event that concerns us, but all possible alternatives. If we are rolling a die, for instance, then the sample space is 1, 2, 3, 4, 5, 6. For the lottery, the sample space is the set of all sequences of six different numbers between 1 and 49. A numerical value is assigned to each event in the sample space, called its 'probability', and this corresponds to how likely that event is to happen. For fair dice each value is equally likely, with a probability of 1/6. Ditto for the lottery, but now with a probability of

1/13,983,816.

We can use a sample space approach to get a ball-park estimate of how amazing the Formula One coincidence was. Top drivers all lap at very nearly the same speed, so the three fastest times can easily fall inside the same tenth-of-a-second period. At intervals of a thousandth of a second, there are one hundred possible lap times for each to 'choose' from: this list determines the sample space. The probability of the coincidence turns out to be one chance in ten thousand. Unlikely enough to be striking, but not *so* unlikely that we ought to feel amazed.

Estimates like this help to explain astounding coincidences reported in newspapers, such as a bridge player getting a 'perfect hand' – all thirteen cards in one suit. The number of games of bridge played every week worldwide is *huge* – so huge that every few weeks the actual events explore a large fraction of the sample space. So occasionally a perfect hand actually does turn up – with the frequency that its small but non-zero probability predicts. The probability of *all four* players getting a perfect hand at the same time, though, is so micoscopic that even if every planet in the galaxy had a billion inhabitants, all playing bridge every day for a billion years, you wouldn't expect it to happen.

Nevertheless, every so often the newspapers report a four-way perfect hand. The sensible conclusion is *not* that a miracle happened, but that *something* changed the odds. Possibly the players got *close* to a four-way perfect hand, and the tale grew in the telling, so that when the journalist arrived with a photographer, another kind of narrative imperative ensured that their story fitted what the journalist had been told. Possibly they deliberately cheated to get their names in the papers. Scientists, especially, tend to underestimate the propensity of people to *lie*. More than one scientist has been fooled into accepting apparent evidence of extrasensory perception or other 'supernatural' events, which can actually be traced to deliberate trickery.

Many other apparent coincidences, on close investigation, slither into a grey area in which trickery is strongly suspected, but may never be proved – either because sufficient evidence is unobtainable, or because it's not worth the trouble. Another way to be fooled about a coincidence is to be unaware of hidden constraints that limit the sample space. That 'perfect hand' could perhaps be explained by the way bridge players often shuffle cards for the next deal, which can be summed up as: poorly. If a pack of cards is arranged so that the top four cards consist of one from each suit, and thereafter every fourth card is in the same suit, then you can cut (but not shuffle, admittedly) the pack as many times as you like, and it will deal out a four-way perfect hand. At the end of a game, the cards lie on the table in a fairly ordered manner, not a random one – so it's not so surprising if they possess a degree of structure after they've been picked up.

So even with a mathematically tidy example like bridge, the choice of the 'right' sample space is not entirely straightforward. The actual sample space is 'packs of cards of the kind that bridge players habitually assemble after concluding a game', *not* 'all possible packs of cards'. That changes the odds.

Unfortunately, statisticians tend to work with the 'obvious' sample space. For that question about Israeli fighter pilots, for instance, they would naturally take the sample space to be all children of Israeli fighter pilots. But that might well be the wrong choice, as the next tale illustrates.

According to Scandinavian folklore, King Olaf of Norway was in dispute with the King of Sweden about ownership of an island, and they agreed to throw dice for it: two dice, highest total wins. The Swedish king threw a double-six. 'You may as well give up now,' he declared in triumph. Undeterred, Olaf threw the dice ... One turned up six ... the other *split in half*, one face showing a six and the other a one. 'Thirteen, I win,' said Olaf.*

Something similar occurs in *The Colour of Magic*, where several gods are playing dice to decide certain events on the Discworld:

> The Lady nodded slightly. She picked up the dice-cup and held it steady as a rock, yet all the Gods could hear the three

* Possibly he was holding a large axe at the time.

cubes rattling about inside. And then she sent them bouncing across the table.

A six. A three. A five.

Something was happening to the five, however. Battered by the chance collision of several billion molecules, the die flipped onto a point, spun gently and came down a seven.

Blind Io picked up the cube and counted the sides.

'Come *on*,' he said wearily. 'Play fair.'

Nature's sample space is often bigger than a conventional statistician would expect. Sample spaces are a human way to model reality: they do not capture all of it. And when it comes to estimating significance, a different choice of sample space can completely change our estimates of probabilities. The reason for this is an extremely important factor – 'selective reporting', which is a type of narrativium in action. This factor tends to be ignored in most conventional statistics. That perfect hand at bridge, for instance, is far more likely to make it to the local or even national press than an imperfect one. How often do you see the headline BRIDGE PLAYER GETS ENTIRELY ORDINARY HAND, for instance? The human brain is an irrepressible pattern-seeking device, and it seizes on certain events that it considers significant, whether or not they really are. In so doing, it ignores all the 'neighbouring' events that would help it judge how likely or unlikely the perceived coincidence actually is.

Selective reporting affects the significance of those Formula One times. If it hadn't been them, maybe the tennis scores in the US Open would have contained some unusual pattern, or the football results, or the golf … Any one of those would have been reported, too – but none of the failed coincidences, the ones that *didn't* happen, would have hit the headlines. FORMULA ONE DRIVERS RECORD DIFFERENT LAP TIMES … If we include just ten major sporting events in our list of would-be's that weren't, that one in ten thousand chance comes down to only one in a thousand.

Having understood this, let's go back to the Israeli fighter pilots. Conventional statistics would set up the obvious sample space, assign probabilities to boy and girl children, and calculate the chance of getting 84% girls in a purely random trial. If this were less than one in a hundred, say, then the data would be declared 'significant at the 99% level'. But this analysis ignores selective reporting. Why did we look at the sexes of Israeli fighter pilots' children in the first place? Because our attention had *already* been drawn to a clump. If instead the clump had been the heights of the children of Israeli aircraft manufacturers, or the musical abilities of the wives of Israeli air traffic controllers, then our clump-seeking brains would again have drawn the fact to our attention. So our computation of the significance level tacitly excludes many other factors that *didn't* clump – making it fallacious.

The human brain filters vast quantities of data, seeking things that appear unusual, and only then does it send out a conscious signal: *Wow! Look at that!* The wider we cast our pattern-seeking net, the more likely it is to catch a clump. For this reason, it's illegitimate to include the data that brought the clump to our attention as part of the evidence that the same clump is unusual. It would be like sorting through a pack of cards until you found the ace of spades, putting it on the table, and then claiming miraculous powers that unerringly accomplish a feat whose probability is one in 52.

Exactly this error was made in early experiments on extra-sensory perception. Thousands of subjects were asked to guess cards from a special pack of five symbols. Anyone whose success rate was above average was invited back, while the others were sent home. After this had gone on for several weeks, the survivors all had an amazing record of success! Then these 'good guessers' were tested some more. Strangely, as time went on, their success rate slowly dropped back towards the average, as if their powers were 'running down'. Actually, that effect wasn't strange at all. It happened because the initial high scores were included in the running total. If they had been omitted, then the scoring rate would have dropped, immediately, to near average.

So it is with the fighter pilots. The curious figures that drew researchers' attention to these particular effects may well have been the result of selective reporting, or selective attention. If so, then we can make a simple prediction: 'From now on, the figures will revert to fifty–fifty.' If this prediction *fails*, and if the results instead confirm the bias that revealed the clump, then the new data can be considered significant, and a significance level can sensibly be assigned by the usual methods. But the smart money is on a fifty–fifty split.

The alleged decline in the human sperm count may be an example of selective reporting. The story, widely repeated in the press, is that over the past 50 years the human sperm count for 'normal' men has halved. We don't mean selective reporting by the people who published the first evidence – they took pains to avoid all the sources of bias that they could think of. The 'selective reporting' was done by researchers who had contrary evidence but didn't publish it because they thought it must be wrong, by journal referees who accepted papers that confirmed a decline more often than they accepted those that didn't, and by the press – who strung together a whole pile of sex-related defects in various parts of the animal kingdom into a single seamless story, unaware that each individual instance has an entirely reasonable explanation that has nothing to do with falling sperm-counts and often nothing to do with sex.

Sexual abnormalities in fish near sewer outlets, for instance, are probably due to excess nitrites, which all fish-breeders know cause abnormalities of *all* kinds – and not to oestrogen-like compounds in the water, which would bolster the 'sperm count' story. Current data from fertility clinics, by the way, show no signs of a decline.

Humans add narrativium to their world. They insist in interpreting the universe as if it's telling a story. This leads them to focus on facts that fit the story, while ignoring those that don't. But we mustn't let the coincidence, the clump, choose the sample space – when we do that, we're ignoring the surrounding space of near-coincidences.

Jack and Ian managed to test this theory on a trip to Sweden. On the plane, Jack predicted that a coincidence would happen at Stockholm airport – for reasons of selective reporting. If they looked hard enough, they'd find one. They got to the bus stop outside the terminal, and no coincidences had occurred. But they couldn't find the right bus, so Jack went back to the enquiries desk. As he waited, someone came up next to him – Stefano, a mathematician who normally occupied the office next door to Jack's. Prediction confirmed. But what was really needed was evidence of a near-coincidence – one that hadn't happened, but could have been selectively reported if it had. For instance, if some other acquaintance had shown up at exactly the same time, but on the wrong day, or at the wrong airport, they'd never have noticed. Near coincidences, by definition, are hard to observe … but not impossible. Ian happened to mention all of the above to his friend Ted, who was visiting soon after. 'Stockholm?' said Ted: 'When?' Ian told him. 'Which hotel?' Ian told him 'Funny, I was staying there *one day later than you*!' Had the trip been one day later, the 'coincidental' encounter with Stefano wouldn't have happened – but the one with Ted would.

What we must not do, then, is to look back at past events and find significance in the inevitable few that look odd. That is the way of the pyramidologists and the tea-leaf readers. Every pattern of raindrops on the pavement is unique. We're not saying that if one such patterns happens to spell your name, this is not to be wondered at – but if your name had been written on the pavement in Beijing during the Ming dynasty, at midnight, nobody would have noticed. We should not look at *past history* when assessing significance: we should look at all the other things that might have happened instead.

Every event is unique. Until we place that event in a category, we can't work out which background to view it against. Until we choose a background, we can't estimate the event's probability. If we consider the sample space of all possible DNA codes, for instance, then we can calculate the probability of a human being having exactly *your* DNA code – which is vanishingly small. But it would be silly to conclude that it is impossible for you to exist.

THIRTY-FIVE

STILL
BLOODY LIZARDS

'THE FUTURE IS LIZARD,' said Ridcully. 'Obviously.'

It was a few days later. The omniscope was focused on a mound of leaves and rotting vegetation a little way from the banks of the river. There was a large depression hanging over the Senior Wrangler, and the Dean had a black eye. The war between land and sea had just entered a terminal stage.

'Little portable seas,' said Ponder. 'You know, I never thought of them like that.'

'An egg is an egg, however you look at it,' said Ridcully. 'Look, you two, I don't want to see a scuffle like that again, d'you hear?'

The Senior Wrangler dabbed at his bleeding nose.

'He goaded be,' he said. 'Id's still osuns, howeber you look at id.'

'A private ocean full of food,' said Ponder, still entranced. 'Hidden in a heap of … well, compost. Which heats up. That's like having private sunshine.'

The little lizard-like creatures that had hatched from the eggs in the mound slithered and slid down the bank into the water, bright-eyed and hopeful. The first few were instantly snapped up by a large male lying in wait among the weeds.

'However, the mothers still have something to learn about post-natal care,' said Ridcully. 'I wonder if they'll have time to learn? And how did they know how to do this? Who's telling them?'

The wizards were depressed again. Most days started that way now. Creatures seemed to turn up in the world randomly, and certainly not according to any pictures in a book. If things were changing into other things, and no one had seen that happen yet, why were the original things still the original things? If the land was so great, why were any fish left in the sea?

The air-breathing fishes that Rincewind had seen still seemed to be around, lurking in swamps and muddy beaches. Things changed, but still stayed the same.

And if there was any truth at all in Ponder's tentative theory that things *did* change into other things, it led to the depressing thought that, well, the world was filling up with quitters, creatures which – instead of staying where they were, and really making a *go* of life in the ocean or the swamp or wherever – were running away to lurk in some niche and grow legs. The kind of fish that'd come out of water was, frankly, a disgrace to the species. It kept *coughing* all the time, like someone who'd just given up smoking.

And there was no purpose, Ridcully kept saying. Life was on land. According to the book, there should be some big lizards. But nothing seemed to be making much of an effort. The moment anything felt safe, it stopped bothering.

Rincewind, currently relaxing on a rock, rather liked it. There were large animals snuffling around in the greenery near the rock he was sitting on; in general shape and appearance, they looked like a small skinny hippopotamus designed in the dark by a complete amateur. They were hairy. They coughed, too.

Things that were doing sufficiently beetle-like things for him to think of them as beetles ambled across the ground.

Ponder had told him the continents were moving again, so he kept a firm grip on his rock just in case.

Best of all, nothing seemed to be *thinking*. Rincewind was convinced that no good came of that sort of thing.

The last few weeks of Discworld time had been instructive. The wizards had tentatively identified several dozen embryo civilizations, or at least creatures that seemed to be concerned about more than simply where their next meal was coming from. And where were they now? There was a squid one, HEX said, out in the really deep cold water. Apart from that, ice or fire or both at once came to the thinkers and the stupid alike. There was probably some kind of moral involved.

The air shimmered, and half a dozen ghostly figures appeared in front of him.

There were, in pale shadowy colours, the wizards. Silvery lines flickered across their bodies and, periodically, they flickered.

'Now, remember,' said Ponder Stibbons, and his voice sounded muffled, 'You are in *fact* still in the High Energy Magic building. If you walk slowly HEX will try to adjust your feet to local ground level. You'll have a limited ability to move things, although HEX will do the actual work –'

'Can we eat?' said the Senior Wrangler.

'No, sir. Your mouth isn't *here*.'

'Well, then, what am I talking out of?'

'Could be anyone's guess, sir,' said Ponder diplomatically. 'We can hear you because our ears are in the HEM, and you can hear the sounds made *here* because HEX is presenting you with an analogue of them. Don't worry about it. It'll seem quite natural after a while.'

The ghost of the Dean kicked at the soil. A fraction of a second later, a little heap of earth splashed up.

'Amazing!' he said, happily.

'Excuse me?' said Rincewind.

They turned.

'Oh, Rincewind,' said Ridcully, as one might say 'oh dear, it's raining'. 'It's you.'

'Yes, sir.'

'Mister Stibbons here's found a way of getting HEX to operate more than one virtually-there suit, d'you see? So we thought we'd come down and smell the roses.'

'Not for several hundred million years, sir,' said Ponder.

'Dull, isn't it,' said the Lecturer in Recent Runes, looking around. 'Not a lot going on. Lots of life, but it's just hanging around.'

Ridcully rubbed his hands together.

'Well, we're going to liven it up,' he said. 'We're going to move things forward fast while we're *here*. A few prods in the right place, that's what these creatures need.'

'The time travelling is not much fun,' said Rincewind. 'You tend to end up under a volcano or at the bottom of the sea.'

'We shall see,' said Ridcully firmly. 'I've had enough of this.

Look at those damn sloppy things over there. 'He cupped his hands and shouted, 'Life in the sea not good enough for you, eh? Skiving off, eh? Got a note from your mother, have you?' He lowered his hands. 'All right, Mister Stibbons ... tell HEX to take us forward, oh, fifty million years – hang on, what was that?'

Thunder rolled around the horizon.

'Probably just another snowball landing,' said Rincewind morosely. 'There's generally one around just when things are settled. It was in the sea, I expect. Stand by for the tidal wave.' He nodded at the browsing creatures, who had glanced up briefly.

'The Dean thinks all this hammering from rocks is making the life on this world very resilient,' said Ridcully.

'Well, that's certainly a point of view,' said Rincewind. 'But in a little while a wave the size of the University is going to wash this beach on to the top of those mountains over there. Then I expect the local volcanoes will all let go ... *again* ... so stand by for a country-sized sea of lava coming the other way. After that there'll probably be outbreaks of rain that you could use to etch copper, followed by a bit of a cold spell for a few years and some fog you could cut up in lumps.' He sniffed. 'That which does not kill you can give you a really bad headache.'

He glanced at the sky. Strange lightning was flickering between the clouds, and now there was a glow on the horizon.

'Damn,' he said, in the same tone of voice. 'This is going to be one of the times when the atmosphere catches fire. I hate it when that happens.'

Ridcully gave him a long blank look, and then said, 'Mister Stibbons?'

'Archchancellor?'

'Make that seventy thousand years, will you? And, er right now, if you would be so good.'

The wizards vanished.

All the insects stopped buzzing in the bushes.

The hairy lizards carried on placidly eating the leaves. Then, something made them look up –

The sun jerked across the sky, became very briefly a reddish-yellow band across a twilight hemisphere, and then the world was simply a grey mist. Below Rincewind's feet it was quite dark, and above him it was almost white. Around him, the greyness flickered.

'Is this what it always looks like?' said the Dean.

'Something has to stand still for a couple of thousand years before you see it at all,' said Rincewind.

'I thought it would be more exciting –'

The light flickered, and sun exploded into the sky, the wizards saw waves around them for a moment, and then there was darkness.

'I told you,' said Rincewind. 'We're under water.'

'The land sank under all the volcanoes?' said Ridcully.

'Probably just moved away,' said Rincewind. 'There's a lot of that sort of thing down here.'

They rose above the surface as HEX adjusted to the new conditions. A landmass was smeared on the horizon, under a bank of cloud.

'See?' said Rincewind. 'It's a pain. Time travel always means you end up walking.'

'HEX, move us to the nearest land, please. Inland about ten miles,' said Ponder.

'You mean I could have just asked?' said Rincewind. 'All this time, I needn't have been walking?'

'Oh, yes.'

The landscape blurred for a second.

'You could have said,' said Rincewind accusingly, as they were rushed past, and sometimes *through*, a forest of giant ferns.

The view stabilized. The wizards had been brought to the edge of the forest. Low-growing shrubs stretched away towards more ferns.

'Not much about,' said Ridcully, leaning against a trunk. 'Can I smoke my pipe here, Stibbons?'

'Since technically you'll be smoking in the High Energy Building, yes, sir.'

Rincewind apparently struck a match on the tree trunk. 'Amazing,' he said.

'That's odd, sir,' said Ponder. 'I didn't think there would be any proper trees yet.'

'Well, here they are,' said Ridcully. 'And I can see at least another three more ...'

Rincewind had already started to run. The fact that nothing can harm you is no reason for not being scared. An expert can *always* find a reason for being scared.

The fact that the nearest trunk had toenails was a good one.

From among the ferns above, a large head appeared on the end of too much neck.

'Ah,' said Ridcully calmly. 'Still bloody lizards, I see.'

THIRTY-SIX

RUNNING FROM DINOSAURS

 WALKING WITH DINOSAURS was a very popular television series in the UK in 2000-2001, and it was equally popular on US television soon afterwards. It portrayed dozens of kinds of dinosaur, all beautifully realised as computer animations in the most intimate detail. And it told us things like 'These doofersaurs were herbivorous, and they were brightly coloured to break up their outline, as protection against predation by thingumosaurs. They were monogamous, rearing their well-protected offspring by living in caves and permitting the children only very restricted access to computer games.'

All this from just two fossil bones – one for each beast.

Walking with Dinosaurs was the latest in a series of popularisations of dinosaurs, from H.G. Wells's *Outline of History* and Arthur Conan Doyle's *The Lost World* to Walt Disney's *Fantasia* and Michael Crichton's *Jurassic Park*. Dinosaurs are redolent with mana; they ooze charisma. They are the PR person's wildest dream. What gives dinosaurs such potency?

The psychologist Helen Haste has taken over from palaeontologist Beverley Halstead the theme of dinosaurs as icons of power and sex in our civilised myth-making. Both have shown that by mythologising them into such potent symbols, we have made it very difficult to work out what it must really have been like when dinosaurs walked the Earth. Our mental images of dinosaurs carry a lot of excess baggage, and it's hard to get away from the lies-to-children involved in those images.

We'll do our best, nonetheless. Whatever it was like, we're sure that *Running from Dinosaurs* is a more apt image.

We have a good idea of what it *wasn't* like. We've all seen the standard filmic shot of dinosaur 'ecology', and it wasn't like that.

The camera homes in on some archaic-looking trees, and we discover a clearing with a lake and some enormous vegetarian reptiles. A few little birdy creatures get into the picture doing their little birdy things; there are pterodactyls flying about ... Then there's a great roar, and a tyrannosaur comes crashing in from stage left. He (nobody can believe it's a 'she' as portrayed: we have to be informed of that in the *Jurassic Park* sequel) jumps on to a brontosaur or a hadrosaur or whatever, and wrestles it to the ground. Or he has a great teeth-against-horns fight with a triceratops, or some other armoured vegetarian like the stegosaur in Disney's *Fantasia*.

In the natural history books of the late nineteenth and early twentieth centuries, the drawings always show these creatures in mortal combat, too (labelled, for example 'Ideal Scene of the Middle Oolitic Period'). Wells's *Outline of History* was atypical in this regard; the drawings and Plates showed the animals simply going about their business, without the drama.

What was it really like?

For a start, there wasn't an 'it'. Dinosaurs, and other great reptiles, had nearly two hundred million years of being the most interesting, most important – well, the biggest, at least – land animals, and they re-invaded the sea and even produced the largest flying creatures that have ever been. Dinosaurs began, about two hundred and forty million years ago, with a single species. The earliest descendants of which we have fossils are the herbivore *Pisanosaurus*, about three feet (1m) long, and *Eoraptor*, a predator of similar size, which date from about 230 million years ago.

By 215 million years ago the dinosaurs had diversified considerably. There were big sprawling amphibians, heavily-built and three or four feet long, as well as a great variety of little ones looking more like big salamanders. There were the synapsid reptiles, which were bigger and held up by sturdy legs: some had great bony sails on their backs, some of them were vegetarians as big as donkeys, some were carnivores as big as hyaenas. There were many kinds of more active dog-sized beasties.

Their therapsid descendants would be the mammals' ancestors, little guys called morganucodontids, which we talk about later. These were small beasties, because most of the action by then was dinosaur. For 150 million years, between 215 and 65 million years ago, any land animal more than three feet (1m) long was a dinosaur.

Among these mammal-like reptiles in the forests were a few active predators the size of kangaroos and wallabies. These unprepossessing creatures were the earliest dinosaurs. You wouldn't have marked them out as having a great future; they were just part of the land fauna that came out of the dark wet Carboniferous forests and began to make a living in the drier Permian era. If you had walked about in those forests you would probably have seen some of these beasts; they were rather slow and stupid, and perhaps they would have attacked you in a rather leisurely, crocodilian way. They weren't as bright as crocodiles, though, nor as quick.

At least as important as the dinosaurs was something far less filmic: soil. By then, soil had developed much of its complex ecosystem: at least fifty interacting species of bacteria, several very different kinds of fungi, insects, worms, protozoans during the Carboniferous. Soil had become a great basis for land plant growth. These forests were not starved, like the modern rainforests with their six inches of soil and no net oxygen production. Oh, no. The coal that we burn for heat and power was deposited in the Carboniferous, and every ton of coal that was deposited by the forests released more than twice its weight of oxygen into the air. Just as each ton of coal we burn now uses that much oxygen to make carbon dioxide again.

In those days, plants grew fast, like modern herbs, but many of them grew big, as well. These plants didn't have woody trunks, they were like great tree-ferns, and the creatures that fed on them made a big input into the terrestrial ecology. Until grass really took over, well after the demise of the dinosaurs, and made savannah and pampas the vegetable basis of the great herds of mammalian herbivores, this thick-soiled forest was the nutritional basis of all land fauna. And the ancient land fauna diversified in these forests, at their edges and in the swamps that they so often became.

There were several early monitor-lizard-like carnivores in the late Carboniferous and early Permian, and their relatives came down to us as the lizards and the snakes. But to get a better idea of what these early reptiles really looked like, and behaved like, you should look at the tuatara, a genuine 'living fossil' from the islands off New Zealand's coast. You can probably find it in your local zoo. It's slow, it's stupid, and any modern iguana or monitor can run rings round it (partly because it's adapted to colder climes); but it's a warning not to use any geckos or pythons or goannas you have known to inform yourself about early Permian lizards.

Slow it may have been, but this lineage became extremely diverse. Its adaptive radiation, its explosive evolutionary diversification, simply swamped the other reptiles, turtles and mammal-like creatures. The early reptiles produced several kinds of marine lizards, of which plesiosaurs and ichthyosaurs were biggest and the most famous. However, another reptile lineage had slid back into the seas in the early Permian, creatures called mesosaurs that were related to turtles and probably lived on plankton that they sieved from the water, like many whales do now. The plesiosaurs, particularly some rather nasty short-necked crocodile-like ones called pliosaurs, were worthy opponents of the big sharks, and probably fed on mesosaurs. But the most successful marine reptiles, as fully adapted to marine life as whales and dolphins are now, were the ichthyosaurs. They flourished long before the famous dinosaurs on land, reaching their peak of size in the Triassic, as far in the past of the tyrannosaur as he was in our past. A length of 30 feet (10m) was common, and occasionally they reached 45 feet (15m).

They were trumped in the seas by a later branch of the lizard lineage, the so-called fish-lizards or mosasaurs, which took over the seas just when the big brontosaurs and allosaurs were taking over the land. Some of them were only a foot (30cm) long, a few stretched to forty feet (12m). But all those films you've seen showing ichthyosaurs in the seas, tyrannosaurs on land are scientifically as (in)accurate as *One Million Years BC* with Raquel Welch being chased by dinosaurian monsters, the Flintstone family with their tame dinosaurs powering household gadgets, or Hamlet with a PC.

It's quite difficult to get the span of geological time to make sense. In his book *In Search of Deep Time*, Henry Gee does an excellent job of reminding us just how flimsy the 'fossil record' is. A few bones here; a few others five thousand miles away and ten million years later; from these we attempt to tell a story of evolutionary ancestry. It's like claiming to have reconstructed human history from one flint flake and a half-eaten hamburger. Well, not as coherent as that, actually.

It's more difficult still to put the range of creatures up against the tapestry of evolutionary time to get some idea of what the dinosaurian Earth might have been like. Take sharks. There have been shark-like sharks since way before there were *any* reptiles; there have been those odd-looking horseshoe crabs for longer. Coelacanth fishes have been grubbing about in the dark depths off the continental shelves while the dinosaurs came and went. It's not surprising, perhaps, that yeasts and other fungi, bacteria of several modern kinds, have been around for more than a thousand million years. We don't expect 'blobs' to notice the passage of time. But coelacanths and sharks and tuataras are vertebrates – you'd think they'd be a bit progressive, evolving and changing into … whatever. But they didn't; they just kept on doing their own thing.

Sharks have eaten mesosaurs, have been annoyed by plesiosaurs and ichthyosaurs and cautious about pliosaurs, have eaten little mosasaurs and been eaten by big ones. There were ammonites in their seas, and belemnites and all kinds of other shelled octopuses. Then, when the big reptiles went, and all the ammonites and their friends disappeared, the sharks had the top levels of the marine food chains to themselves for tens of thousands of years. Then the mammals produced dolphins, killer whales, big whales … and the sharks just went on being sharks.

Why didn't sharks change? They have wonderful immune systems, which we're just beginning to understand; they don't get bacterial infections or, apparently, cancers. Perhaps they are not made ill by viruses, either. Though they do have lots of wormy and flukey parasites. Are today's sharky superfish the newest, latest-model sharks? Or did the ancient ones also have marvellous immunity to disease? Was that their trick? Was that what brought them unchanged, in outward appearance at least, through such a long span of time?

Since we can't answer those questions, yet, let's move to one that we can answer. What was happening on land as the Permian became the Triassic? Well, for a start, the biggest mass extinction of all, 248 million years ago. The number of species on Earth dropped by about 93%. Only 7% survived. We're talking big species here; nobody knows what happened to bacteria or even protozoa, to nematode worms, or to rotifers. Except that a lot of protozoa have mineral shells, and got fossilised as the white cliffs of Dover, and rotifers have tiny hard jaws of characteristic shapes that a few hardy fossil-hunters collect. So we can check these out, and they give a similar picture.

The precise cause of this mass extinction is debatable. There may have been collusion between a comet impact and massive volcanic activity, as we describe shortly. In any event, the early dinosaurs – and mammal-like reptiles and turtles, even ichthyosaurs and early plesiosaurs (but not mesosaurs) – lived through it. They were among the lucky 7%. Whatever the disaster was, it seemed great for them; it gave them just the open ecological spaces that they needed to radiate wildly. The Triassic seas were just as full of reptiles as today's seas are full of mammals, and they stayed that way right up to the early Cretaceous. Those ocean reptiles had largely vanished before *Tyrannosaurus* appeared, though.

Why were the reptiles so successful as sea creatures? There's a persuasive biological explanation for this invasion of the seas by land animals. The story starts in the sea, moves on to land, and reverts to the sea again.

Creatures that live in the seas experience very little gravity, if any; even heavily-armoured creatures like crabs produce forms that swim. In fact their muscles are for swimming, or for closing jaws on other creatures, or for sudden escapes. But as the descendants of these creatures came out on land, they experienced real problems; they sagged and flopped. Compare a salamander with its sprawled-out legs, which cannot support its weight, with a lizard of the same size that has strong enough muscles and bones to run. (The tuatara still sprawls). The frog's trick, one jump at a time, is much less effective than really well-designed legs, strong limb girdles, an effective bellows system around the lungs to supply the muscles with oxygen, and a four-chambered heart to keep the aerated and depleted blood apart – all the tricks that make a monitor lizard such a good predator.

When you've achieved all these, then feeding on the seashore, or even in the sea like today's marine iguanas on the Galápagos, becomes really easy. Instead of having muscles and respiratory systems that are just adequate for marine life, you've got the supercharged version that terrestrial gravity required of your ancestors. Going back into the sea is now a very viable option; only sharks and octopuses are any match for previously-terrestrial lineages. So the ichthyosaurs and plesiosaurs, like today's dolphins and whales (but less so because mammals had become warm-blooded too while they were on land) found that living in the sea was easy.

Until they evolved further and became their own best enemies, of course, after they had radiated into hundreds of species. The reptile-eating pliosaurs were like today's killer whales, whose main diet is other whales.

Meanwhile, several rather different lineages of Triassic land reptiles, somewhat prematurely called archosaurs ('ruling reptiles'), had evolved really good limb girdles, of several patterns. Two very different lineages, 'bird-hipped' and 'lizard-hipped', went on to produce all those great big dinosaurs. The lizardy lot evolved into gigantic herbivores like *Diplodocus* and brontosaurs (now, alas, named apatosaurs; 'thunder-lizard' was so much more appropriate) and gigantic carnivores like the allosaurs and the tyrannosaurs. These all arrived at least sixty million years later, though: the early archosaurs were as far in the past of tyrannosaurs as we are in their future.

The bird-hipped lot eventually produced those spectacularly armoured beasts that make such good film shots when tyrannosaurs are fighting them: ankylosaurs with spikes on a knobby tail, like that spiky ball on a chain that the villain wields in chivalric films;

stegosaurs with bony plates, spikes down their backs; triceratops with the bony frill and three horns.

Filmmakers always seem to make almost-purposeful mistakes: the innumerable eight-year old boys who've learned all the lovely names could correct them. It's a pity that that otherwise lovely fight in Disney's *Fantasia*, accompanied by Stravinsky's *Rite of Spring*, is between a tyrannosaur and stegosaur. It couldn't have happened, they weren't contemporary. And the stegosaur never had the anky-losaur's tail-armour, either. The late Cretaceous landscape, with the few very big dinosaurs that lived then, was doubtless an impressive scene; but the film producers' version is no more accurate than *Fantasia*.

This is hardly a surprise; after all, Hollywood has been wrong about so many other things. Scientists don't always fare better. Right now, palaeontologists believe that tyrannosaurs were scav-engers, not predators. We'll stick our necks out and dispute that conclusion. Yes, tyrannosaurs may not have been fearsome preda-tors, hunters ... but if they weren't, that doesn't make scavenging the only option. They probably did something that we can't imag-ine instead. We simply can't see these animals as enormous vultures, with their tiny front feet scrabbling at a decaying corpse and that great head hidden in the abdomen of a dead sauropod like *Diplodocus*. We'd run away from them, whatever the scientists are saying now.

On the Rincewind principle, you understand. Just in case.

Other archosaurs went on to produce crocodiles and ptero-dactyls, and perhaps the birds – or these may have arisen from the same stock as those deinonychids made familiar by *Jurassic Park*, namely *Velociraptor* and its ilk. These probably were intelligent creatures, agile carnivores much like their portrayal in *Jurassic Park*. We'd certainly run away from *them*.

There are a few puzzles that biologists keep puzzling at, like whether some of the big dinosaurs were warm-blooded. And why everything was so damned big in the Cretaceous. The biggest teleost (bony) fish that there has ever been lived in Cretaceous seas;

it was as big as today's whale shark. And dinosaurs flew, too. There are lots of fossil pterosaurs like *Pteranodon* with eight-yard (7m) wingspreads, bigger than any of today's birds, and there are a couple of fossils of *Quetzalcoatlus* with wingspreads twice as big. That's bigger than the fighter-planes of World War II, like spitfires. We have no idea how these creatures lived, but there's no doubt that they existed.* Unless you believe in planetary engineers with a sense of humour, as in *Strata* ...

This is where we should warn you about the most tempting way of thinking about these ancient creatures, and how we get it wrong all the time. Gee's *In Search of Deep Time* takes us to task, showing how all of our pretty guesses about evolutionary scenarios – however well they seem to be based in the fossil record – are simply wrong, impossible.

A classical mistake is the way we habitually think about that fish that came out of the water, whose descendants were the land vertebrates. We imagine it flopping out on the beach (and Rincewind encouraging it to go back into the water) with its developing lungs and its evolving legs ... No. It had fairly well-developed legs while it was still living an underwater life, with internal gills like any perfectly respectable fish. It must have done, or it wouldn't have *worked*.

We have little idea what its legs were for at that stage, though: certainly not for walking on land. But there's no question that these hands that we're typing with are the descendants of those fishy legs ... Just as our cough is a legacy of that fish's crossover of its airway and foodway. It's the pictures in our heads, of what we suppose happened, that are mistaken. They are lies-to-children that we can't correct yet. But humans definitely evolved from a fish, and it had legs. It just didn't use them to walk along the beach.

Another interestingly wrong lie-to-children about evolution

---

* Some pterodactyls were sparrow-sized, and their wings could have carried them easily; but pteranodons can't have taken off from the sea surface, or climbed up cliffs to hang-glide! Although *Quetzalcoatlus* has been modelled, and flown as a radio-controlled glider, that doesn't tell us how it lived. Was there perhaps aerial plankton for it to eat?

concerns the origin of birds. When those lovely *Archaeopteryx* fossils were found in the Solenhofen limestone, so well preserved that the feather imprints could be clearly seen – and the teeth and the claw-digits on the wings – it was clear that we had found the halfway stage between a reptile and a bird. It was a super Missing Link.

Think about that for a moment: how can you find a Missing Link? Oops.

Archaeopteryx had a long tail like a lizard, and no keel in its chest for strong flight muscles. Were it not for the feathers, it would have been classified as a small pseudosuchian dinosaur like *Ornithomimus* (bird-mimic). In the late Jurassic there were many little dinosaurs with bird-like features, and some really well-developed diving-bird fossils had been found in early Cretaceous rocks from about fifteen million years later. These were real birds, called *Ichthyornis*, and they had already lost their flight abilities, reduced their wings to (very bird-like) vestiges.

So *Archaeopteryx* was a bit 'late', and in the 1950's zoologists thought that it probably represented a primitive reptile-bird lineage hanging on, perhaps contemporary with much more bird-like creatures. That's another scenario that doesn't make sense now. Many bird-like dinosaurs have been found recently, fossils from South America and, especially, *Caudipteryx* and *Protoarchaeopteryx* from China, and these make the problems worse.* They have feathers, but *they didn't fly*. They have no wings, but they have arms with hands, sometimes with only two digits.

So what were feathers 'for'?

Feathers are really remarkably complicated. They're not at all like the scales of lizards and snakes, and it's not easy to invent an evolutionary route by which feathers (or hairs, come to that) can have developed from scales. Some biologists who haven't looked at feathers very carefully have imagined scales growing rather like the

---

* Though some palaeontologists think that such creatures were not feathered dinosaurs, but flightless birds. According to John Ruben, *Caudipteryx* was merely 'a Cretaceous turkey, so to speak'.

stage witch's fingernails, so that they stick out like a pangolin's scales. (That's that silly mammalian tree-climbing anteater that looks like a big pine cone.)

Feathers aren't like that, though. They develop as cylinders: you can see baby feathers, called pin-feathers, on a plucked chicken from the supermarket. The scales on birds' legs are respectable reptilian scales; perhaps surprisingly, no modern birds have structures that are halfway between feathers and scales, even though their remote ancestors had scales all over. Probably ... we can't be sure from those very old fossils. Scales are, and probably were in those ancestors, patches of keratin very like fingernails; sometimes they overlap like roof tiles. Feathers are cylinders sitting in follicles, deep pits in the skin. About a millimetre up from their deep end, they have a ring of dividing cells, called the collar, which produce the cylinder by growing it outwards. As the products move up the follicle, they turn all the cells' productivity into making keratin, the protein of horn, nails, hair and feathers. And the cylinder wall becomes horny in a strange pattern.

The side of the feather facing backwards on the bird produces creases that pass around on both sides of the cylinder toward the front-facing side, diving into the follicle so that they are almost parallel with the length of the cylinder. They don't quite meet, and the tissue between their deep ends will become the stalk of the feather. The other side splits open, and the barbs of the feather – between the creases – unfurl to make the feather vane. They are much longer than the cylinder's circumference, so a narrow pin-feather can generate a feather with a broad vane.

Not a bit like a scale. And far more complicated. Evolution had to work hard to come up with feathers.

And they must have evolved for a good reason, because lots of dinosaurs had feathers in various versions. Some were like down-feathers, others more like paintbrushes or feather dusters. They could have evolved for warmth. Adult velociraptors and young tyrannosaurs may have been covered in down, like baby chicks. Palaeontologist Mark Norell says that 'We have as much evidence that velociraptors had feathers as we do that Neanderthals had hair'. But other scientists disagree.

Perhaps feathers were sexual ornaments. More likely, we haven't yet thought of whatever function they had.

So much for stories that 'some reptiles got feathers and became birds'.

There's a very general problem here, and it's the problem the wizards are having all the time. The overall pattern of evolution, of the birds say, or of tyrannosaurs, looks very sensible. But at a deep level, deeper than the 'common-sense story' that the wizards are trying to use for Roundworld, we don't *understand* it. We have stories to explain it, all right, but that isn't what understanding means.

In fact, we're not altogether sure what it does mean, in a scientific sense. We know that apparently quite ordinary historical events 'come to pieces' when you try to analyse them from several directions. John F. Kennedy's assassination is a perfect example: the bullets seem to have come from different directions, and there doesn't appear to be one consistent story that means we can 'understand' what happened. We can describe the events, but the underlying causalities, like the physicists' quantum theory and relativity theory, don't match up.

Evolution doesn't just happen to one creature at a time. The entire ecosystem evolves, and as it does so, new tricks may become worth using, for a limited period of time and in a limited region of geography. A few of those tricks turn out to have useful effects that are quite different from the 'reason' why they evolved to begin with, and those effects may continue to be worth having long after the original reason for their appearance has ceased to apply.

It's not surprising, in that frame, that historical events as far back as the origin of feathers, or of birds, don't make detailed sense either. That's why we can't imagine what it was like in the Late Cretaceous. *Walking with Dinosaurs* was beautifully made and based on up-to-date science, but ultimately it was unconvincing. The need to tell a story distorted what was actually known, and mixed it up with guesswork and wishful thinking. (You can't be sure what colour an animal's skin was when all you've got is fossil bones. And assuming it was a bit like something vaguely similar that's alive today is cheating, not science.) Anything televisual was automatically preferred to some more mundane scenario. So we got a dinosaur soap, with everything over-dramatised.

So we can't imagine ourselves walking with dinosaurs. Neither, to be fair, can we imagine ourselves running from them. We simply can't understand what it was like to be there, much as we'd like to believe we can. Still, what can you expect from a slightly brainy ape? We can give ourselves an excuse, an evolutionary one, and it's this: we're only halfway from *Australopithecus* (Lucy) to human. It's not surprising that we keep finding limits to what our brains can comprehend. Neither is it surprising that our mental models won't fit together to offer a convincing understanding of past events. We're not smart enough. Yet.

Wait another million years. Then we'll show you.

THIRTY-SEVEN

I SAID, DON'T
LOOK UP!

 PONDER WAS WORKING THE RULES AGAIN. Now they read:

THE RULES
1 Things fall apart, but centres hold
2 Everything moves in curves
3 You get balls
4 Big balls tell space to bend
5 There are no turtles anywhere
   (after this one he'd added *Except ordinary ones*)
6 Life turns up everywhere it can
7 Life turns up everywhere it can't
8 There is something like narrativium
9 There may be something called bloodimindium (see rule 7)
10 …

He stopped to think. Behind him, a very large lizard killed and ate a slightly smaller one. Ponder didn't bother to turn around. They'd been watching lizards for more than a hundred million years – all day, in fact – and even the Dean was giving up on them.

'*Too* well adapted,' he said. 'No *pressure* on them, you see.'

'They're certainly very dull,' said Ridcully. 'Interesting colours, though.'

'Brain the size of a walnut and some of them think with their backsides,' said the Senior Wrangler.

'Your type of people, Dean,' said Ridcully.

'I shall choose to ignore that, Archchancellor,' said the Dean coldly.

'You've been interfering again, haven't you,' Ridcully went on. 'I saw you pushing some of the small lizards out of that tree.'

'Well, you've got to admit that they *look* a bit like birds,' said the Dean.

'And did they learn to fly?'

'Not in so many words, no. Not horizontally.'

'Eat, fight, mate and die,' said the Lecturer in Recent Runes. 'Even the crabs were better than this. Even the blobs made an effort. When they come to write the history of this world, this is the page everyone will skip. Terribly dull lizards, they'll be called. You mark my words.'

'They have stayed around for a hundred million years, sir,' said Rincewind, who felt he had to stand up for non-achievers.

'And what have they done? Is there a single line of poetry? A building of any sort? A piece of simple artwork?'

'They've just not died, sir.'

'Not dying out is some kind of achievement, is it?' said the Lecturer in Recent Runes.

'Best kind there is, sir.'

'Pah!' said the Dean. 'All they prove is that species go soft when there's nothing happening! It's nice and warm, there's plenty to eat … it's just the sea without water. A few periods of vulcanism or a medium-sized comet would soon have them sitting up straight and paying attention.'

The air shimmered and Ponder Stibbons appeared.

'We have intelligence, gentlemen,' he said.

'I know,' said the Dean.

'I mean, the omniscope has found signs of developing intelligence. Twice, sir.'

The herd was big. It was made up of large, almost hemispherical creatures, with faces that had all the incisive cogitation of a cow.

Much smaller creatures were trotting along at the edges. They were dark, scrawny and warbled to one another almost without cease.

They also carried pointed sticks.

'Well …' Ridcully began, dismissively.

'They're herding them, sir!' said Ponder.

'But wolves chase sheep ...'

'Not with pointy sticks, sir. And look there ...'

One of the beasts was towing a crude travois, covered with leaves. Several herders were lying on it. They were pale around the muzzles.

'Are they sick, d'you think?' said the Dean.

'Just old, sir.'

'Why'd they want to slow themselves down with a lot of old people?'

Ponder dared a short pause before answering.

'They're the library, sir. I suppose. They can remember things. Places to hunt, good waterholes, that sort of thing. And that means they must have some sort of language.'

'It's a start, I suppose,' said Ridcully.

'Start, sir? They've nearly done it all!' Ponder put his hand to his ear.

'Oh ... and HEX says there's more, sir. Er ... different.'

'How different?'

'In the sea again, sir.'

'Aha,' said the Senior Wrangler.

In fact on the sea was more accurate, he had to admit. The colony they found stretched for miles, linking a chain of small rocky islands and sandbanks as beads on a chain of tethered driftwood and rafts of floating seaweed.

The creatures inhabiting it were another type of lizard. Still extremely dull, the wizards considered, compared to some of the others. They weren't even an interesting colour and they had hardly any spikes. But they were ... busy creatures.

'That seaweed ... does it look sort of *regular* to you?' said the Lecturer in Recent Runes, as they drifted over a crude wall. 'They're not *farming*, are they?'

'I think ...' Ponder looked down. The water washed over the wall of rocks. 'It's a big cage for fish. The whole lagoon. Er ... I think they've built the walls like that so the tide lets the fish come in and then they're stuck when it goes down.'

Lizards turned their heads as the semi-transparent men floated past, but seemed to treat them as no more than passing shadows.

'They're harnessing the power of the sea?' said Ridcully. 'That's clever.'

Lizards were diving at the far side of the lagoon. Some were busy around rock pools on one of the lower islands. Small lizards swam in the shallows. Along one stretch of driftwood walkways, strips of seaweed were drying in the breeze. And over everything was a yip-yipping of conversation. And it *was* conversation, Ponder decided. Animals didn't wait for other animals to finish. Nor did wizards, of course, but they were a breed apart.

A little way away, a lizard was carefully painting the skin of another lizard, using a twig and some pigments in half-shells. The one doing the painting was wearing a necklace of different shells, Ponder realized.

'Tools,' he murmured. 'Symbols. Abstract thought. *Things* of value … is this a civilization, or are we merely tribal at the moment?'

'Where's the sun?' said the Senior Wrangler. 'It's always so hazy, and it's hard to get used to directions here. Wherever you point, it's at the back of your own head.'

Rincewind pointed towards the horizon, where there was a red glow behind the clouds.

'I call it Widdershins,' he said. 'Just like at home.'

'Ah. The sun sets Widdershins.'

'No. It doesn't do anything,' said Rincewind. 'It stays where it is. The horizon comes up.'

'But it doesn't fall on us?'

'It tries to, but the other horizon drags us away before it happens.'

'The more time I spend on this globe, the more I feel I should be holding on to something,' the Dean muttered.

'And the light isn't reflected around the world?' said the Senior Wrangler. 'It is at home. It's always very beautiful, the glow that comes up through the waterfall.'

'No,' said Rincewind. 'It just gets dark, unless the moon is up.'

'And there's still just the one sun, isn't there?' said the Senior Wrangler, a man with something on his mind.

'Yes.'

'We didn't add another one?'

'No.'

'So … er … what is that light over there?'

As one wizard, they turned towards the opposite horizon.

'Whoops,' said the Dean, as the distant thunder died away and lights streamed high across the sky.

The lizards had heard it too. Ponder looked around. They were lining the walkways, watching the horizon with all the intelligent interest of a thinking creature wondering what the future may hold …

'Let's get back to the High Energy Magic building before the boiling rain, shall we?' said Ridcully. 'This really is too depressing.'

THIRTY-EIGHT

THE DEATH OF DINOSAURS

 LIFE TURNS UP EVERYWHERE IT CAN.

Life turns up everywhere it can't.

And just when it seems to have got itself going really comfortably, with a sustainable lifestyle and gradual progress towards higher things, along comes a major catastrophe and sets it back twenty million years. Yet, paradoxically, those same disasters also pave the way to radically new lifeforms ...

It's all rather confusing.

Life is resilient, but any particular species may not be. Life is constantly devising new tricks. The one with eggs is brilliant: provide the developing embryo with its own personal life-support machine. Inside, the environment is tailored to the needs of that species – and what's outside doesn't matter much, because there's a barrier to keep it out.

Life is adaptable. It changes the rules of its own game. As soon as eggs make their appearance, the stage is set for the evolution of egg-eaters ...

Life is diverse. The more players there are, the more ways there are to make a living by taking in each others' washing.

Life is repetitious. When it finds a trick that works, it churns out thousands of variations on the same basic theme. The great biologist John (J.B.S.) Haldane was once asked what question he would like to pose to God, and replied that he'd like to know why He has such an inordinate fondness for beetles.*

There are a third of a million beetle species today – far more than in any other plant or animal group. In 1998 Brian Farrell came

* Readers of the Discworld book *The Last Continent* will recall that, by an amazing coincidence, beetles were something of a passion for the God of Evolution.

up with a possible answer to Haldane's query. Beetles appeared about 250 million years ago, but the number of species didn't explode until about 100 million years ago. That happens to be just when flowering plants came into existence. The 'phase space' available for organisms suddenly acquired a new dimension – a new resource became available for exploitation. The beetles were beautifully poised to take advantage by eating the new plants, especially their leaves. It used to be thought that flowering plants and pollinating insects drove each other to wilder and wilder diversity, but that's not true. However, it *is* true for beetles. Nearly half of today's beetle species are leaf-eaters. It's *still* an effective tactic.

Sometimes natural disasters don't just eliminate a species or two. The fossil record contains a number of 'mass extinctions' in which a substantial proportion of all life on Earth disappeared. The best-known mass extinction is the death of the dinosaurs, 65 million years ago.

In order not to mislead you, we should point out at once that there is no scientific evidence for the existence of any dinosaur *civilization*, no matter what events are going on in the Roundworld Project. But … whenever a scientist says 'there is no scientific evidence for', there are three important questions you should ask – especially if it's a government scientist. These are: 'Is there any evidence *against*?', 'Has anyone looked?', and 'If they did, would they expect to find anything?'*

The answers here are 'no,' 'no', and 'no'. Deep Time hides a lot, especially when it's assisted by continental movement, the bulldozing ice sheets, volcanic action and the occasional doomed asteroid. There are few surviving *human* artefacts more than ten thousand years old, and if we died out today, the only evidence of our civilization that might survive for a million years would be a few dead probes in deep space and various bits of debris on the Moon. *Sixty-*

---

* Rincewind would add some more:
  'Is it safe?'
  'Are you sure?'
  'Are you absolutely sure?'

five million? Not a chance. So although a dinosaurian civilization is pure fantasy – or, rather, pure speculation – we can't rule it out *absolutely*. As for dinosaurs who were sufficiently advanced to use tools, herd other dinosaurs ... well, Deep Time would wash over them without a ripple.

Dinosaurs are always among the most popular exhibits at museums. They remind us that the world wasn't always like it is now; and they remind us that humans have been on this planet for a very short time, geologically speaking. Basically, dinosaurs *are* ancient lizards. The ones whose bones we all go to gawp at in museums are rather *big* lizards, but many were much smaller. The name means 'terrible lizard', and anyone who watched *Jurassic Park* will understand why.

An Italian fossil collector who watched the Spielberg movie suddenly realized that a perplexing fossil, filed away for years in his basement, might well be a bit of a dinosaur. He then sent it to a nearby university, where it was found not just to be a dinosaur, but a new species. It was a young therapod – small flesh-eating dinosaurs that are the closest relatives of birds. Interestingly, it didn't have any feathers. A story straight out of the movies: narrative imperative at work in our own world ... traceable, as always, to selective reporting. How many fossil hunters owned a bit of dinosaur bone but *didn't* make the connection after seeing the movie?

In the human mind, dinosaurs resonate with myths about dragons, common to many cultures and many times; and many miles of suggestions have appeared to explain how the dragon-thoughts in our minds have come down to us, over millions of years of evolution, from real dinosaur images and fears in the minds of our ancient ancestors. However, those ancestors must have been *very* ancient, for those of our ancestors that overlapped the dinosaurs were probably tiny shrewlike creatures that lived in holes and ate insects. After more than a hundred million years of success, the dinosaurs all died out, 65 million years ago – and the evidence is that their demise was sudden. Did proto-shrews have nightmares about dinosaurs, all that time ago? Could such nightmares have survived 65 million years of natural selection? In particular, do shrews today have nightmares about fire-breathing dragons – or is it just us? It seems likely that the dragon myth comes from other, less literal, tendencies of that dark, history-laden organ that we call the human mind.

Dinosaurs exert a timeless fascination, especially for children. Dinosaurs are genuine monsters, they actually existed – and some of them, the ones we all know about, were gigantic. They are also safely dead.

Many small children, even if they are resistant to the standard reading materials in school, can reel off a long list of dinosaur names. 'Velociraptor' was not notable among them before *Jurassic Park*, but it is now. Those of us who still have an affection for the brontosaur often need to be reminded that for silly reasons science has deemed that henceforth that sinuous swamp-dwelling giant must be renamed the apatosaur.* So attuned are we to the dinosaurs that the drama of their sudden disappearance has captured our imaginations more than any other bit of palaeontology. Even our *own* origins attract less media attention.

What about the sudden demise?

For a start, quite a few scientists have disputed that it ever *was* sudden. The fossil record implicates the end of the Cretaceous

---

* A worse case is what used to be called *Eohippus*, the Dawn Horse – a beautiful, poetic name for the animal that formed the main stem of the horse's family tree. It is now called *Hyracotherium*, because somewhat earlier somebody had given that name to a creature that they *thought* was a relative of the hyrax, represented by a single fossil shoulder-blade. Then it turned out that the bone was actually part of an *Eohippus*. Unfortunately, whoever officially names a species *first* must get priority, so now the Dawn Horse has a silly, unpoetic name that commemorates a mistake.

We lost *Brontosaurus* – thunder-lizard – for a similar reason. Thunder Lizard ... what a marvellous name. 'Apatosaurus'? It probably means 'Gravitationally challenged Lizard'.

The moral of this rule is that when learned committees of elderly scientists meet to discuss an exceptional issue they can always be trusted to make a completely ridiculous decision. Quite unlike the wizards of Unseen University, naturally.

period, 65 million years ago, as 'D-Day'. This was also the start of the so-called Tertiary period, or Age of Mammals, so the end of the dinosaurs is usually called the K/T boundary – 'K' because Germans spell Cretaceous with a K. But if we assume that the end of the Cretaceous was 'when it happened', then many species seemed to have anticipated their end by vanishing from the fossil record five to ten million years earlier. Did amorous dinosaurs, perhaps, say to each other 'It's just not worth going through with this reproduction business, dear – we're all going to be wiped out in ten million years.'? No. So why the fuzzy fade-out over millions of years? There are good statistical reasons why we might not be able to locate fossils right up to the end, even if the species concerned were still alive.

To set the remark in context: how many specimens of *Tyrannosaurus rex*, the most famous dinosaur of all, do you think that the world's universities and museums have between them? Not copies, but originals, dug from the rock by palaeontologists?

Hundreds … surely?

No. Until *Jurassic Park*, there were precisely *three*, and the times when those particular animals lived have a spread of five million years. 19 more fossilized *T. rex*es have been found since, probably more by the time you read this, because *Jurassic Park* gave dinosaurs a lot of favourable publicity, making it possible to drum up enough money to go out and find some more. With that rate of success, the chance of a future race finding any fossil humanoids, over the whole period of our and our ancestors' existence, would be negligible. So if some species had survived on Earth for a five million year period, it is entirely likely that *no* fossils of it will have been found – especially if it lived on dry land, where fossils seldom form. This may suggest that the fossil record isn't much use, but quite the contrary applies. Every fossil that we find is proof positive that the corresponding species did actually exist; moreover, we can get a pretty accurate impression of the grand flow of Life from an incomplete sample. One lizard fossil is enough to establish the presence of lizards – even if we've found only one species out of the ten thousand that were around.

Bearing this in mind, though, we can easily see that even if the death of the dinosaurs was extremely sudden, then the fossil record might easily give a different impression. Suppose that fossils of a given species turn up randomly about every five million years. Sometimes they're like buses, and three come along at once – that is, within a million years of each other. Other times, they're also like buses: you wait all day (ten million years) and don't see any at all. During the ten million year run-up to the K/T boundary, you find random fossils. For some species, the last one you find is from 75 million years ago; for others it's from 70 million years ago. For a few, by chance, it's from 65 million years ago. So you *seem* to see a gradual fade-out.

Unfortunately, you'd see much the same if there really *had* been a gradual fade-out. How can you tell the difference? You should look at species whose fossils are far more common. If the demise was a sudden one, those ought to show a sharper cut-off. Species that live wholly or partially in water get fossilized more often, so the best way to time the K/T mass extinction is to look at fossils of marine species. Wise scientists therefore mostly ignore the dinosaur drama and fiddle around with tiny snails and other undramatic species instead. When they do, they find that ichthyosaurs also died out about then, as did the last of the ammonites* and many other marine groups. So something sudden and dramatic really did happen at the actual boundary, but there may well have been a succession of other events just before it too.

What kind of drama? An important clue comes from deposits of iridium, a rare metal in the Earth's crust. Iridium is distinctly more common in some meteorites, particularly those from the asteroid belt between Mars and Jupiter. So if you find an unusually rich deposit of iridium on Earth, then it may well have come from an impacting meteorite.

In 1979 the Nobel-winning physicist Luis Alvarez was musing

* Lots of ammonite species died out 5–10 million years before the K/T boundary, so it looks as if their extinction genuinely was gradual. But whatever it was that happened at the K/T boundary finished them off.

along such lines, and he and his geologist son Walter Alvarez discovered a layer of clay that contains a hundred times as much iridium as normal. It was laid down right at the K/T boundary, and it can be found over the whole of the Earth's land mass. The Alvarezes interpreted this discovery as a strong hint that a meteorite impact caused the K/T extinction. The total amount of iridium in the layer is estimated to be around 200,000 tons (tonnes), which is about the amount you'd expect to find in a meteorite 6 miles (10 km) across. If a meteorite that size were to hit the Earth, travelling at a typical 10 miles per second (16 kps), it would leave an impact crater 40 miles (65 km) in diameter. The blast would have been equivalent to thousands of hydrogen bombs, it would have thrown enormous quantities of dust into the atmosphere, blanking out sunlight for years, and if it happened to hit the ocean – a better than 50/50 chance – it would cause huge tidal waves and a short-lived burst of superheated steam. Plants would die, large plant-eating dinosaurs would run out of food and die too, carnivorous dinosaurs would quickly follow. Insects would on the whole fare a little better, as would insect-eaters.

Much evidence has accumulated that the Chicxulub crater, a buried rock formation in the Yucatan region of southern Mexico, is the remnant of this impact. Crystals of 'shocked' quartz were spread far and wide from the impact site: the biggest ones are found near the crater, and smaller ones are found half way round the world. In 1998 a piece of the actual meteorite, a tenth of an inch (2.5 mm) across, was found in the north Pacific Ocean by Frank Kyte. The fragment looks like part of an asteroid – ruling out a possible alternative, a comet, which might also create a similar crater. According to A. Shukolyukov and G.W. Lugmair, the proportions of chromium isotopes in K/T sediment confirm that view. And Andrew Smith and Charlotte Jeffery have found that mass diebacks of sea urchins which occurred at the K/T boundary were worst in the regions around central America, where we think the meteorite came down.

Although the evidence for an impact is strong, and has grown considerably over the twenty years since the Alvarezes first advanced their meteorite-strike theory, a strongly dissenting group of palaeontologists has looked to terrestrial events, not dramatic astronomical interference, to explain the K/T extinction. There was certainly a rapid series of climatic changes at the end of the Cretaceous, with very drastic changes of sea level as ice caps grew or melted. There is also good evidence that some seas, perhaps all, lost their oxygen-based ecology to become vast, stinking, black, anaerobic sinks. The fossil evidence for this consists of black iron- and sulphur-rich lines in sediments. The most dramatic terrestrial events were undoubtedly associated with the vulcanism which resulted in the so-called Deccan Traps, huge geological deposits of lava. The whole of Asia seems to have been covered with volcanoes, and they produced enough lava that it would have formed a layer 50 yards (45 m) thick if it had been spread over the whole continent. Such extensive vulcanism would have had enormous effects on the atmosphere: carbon dioxide emissions that warmed the atmosphere by the greenhouse effect, sulphur compounds resulting in terrible acid rain and freshwater pollution over the entire planet, and tiny rock particles blocking sunlight and causing 'nuclear winters' for decades at a time. Could the volcanoes that formed the Deccan Traps have killed the dinosaurs, instead of a meteorite? Much depends on the timing.

Our preferred theory, not because there is good independent evidence for it but because it would explain so much, and because it has a moral, is that the two causes are linked. The Chicxulub crater is very nearly opposite the Deccan Traps, on the other side of the planet. Perhaps volcanic activity in Asia began some millions of years before the K/T boundary, causing occasional ecological crises for the larger animals but nothing really final. Then the meteorite hit, causing shockwaves which passed right through the Earth and converged, focused as if by a lens on just that fragile region of the planet's crust. (A similar effect happened on Mercury, where a gigantic impact crater called the Caloris Basin is directly opposite 'weird terrain' caused by focused shockwaves.)

There would then have been a gigantic, synchronized burst of vulcanism – on top of all the events of the collision, which would have been pretty bad on their own. The combination could have polished off innumerable animal species. In support of this idea, it should be said that another geological deposit, the Siberian Traps, contains ten times as much lava as the Deccan system, and it so happens that the Siberian Traps were laid down at the time of another mass extinction, the great Permian extinction, which we mentioned earlier. To pile on further evidence: some geologists believe they have found another meteorite impact site in modern Australia, which in Permian times was opposite to Siberia.

Is there any evidence for this suggestion? In 2000, Dallas Abbott and Ann Isley analysed impact events to see whether they coincided with volcanic 'superplumes', where large amounts of liquid rock well up from the Earth's mantle. A superplume is thought to be responsible for the Hawaiian island chain, for instance, whose relics stretch across half the Pacific Ocean. The likely explanation is that as the continents drifted, the Pacific Plate moved, so that the superplume welled up in what now seem to be different places. However, very recent observations suggest that superplumes can move, too. The Deccan and Siberian Traps were also probably caused by superplumes.

The result of the analysis is that impacts and superplumes occur together far more often than they would if their association is due to chance. That suggests that impacts can make superplumes bigger, or more likely. (It's hard to see how superplumes could make impacts more likely, except by attracting invading aliens who are heavily into vulcanism.) However, thanks to continental drift, it is not certain that the Deccan Traps were in fact on the exact far side of the Earth from Chicxulub at the time when the K/T meteorite hit. So we can't yet be sure that the meteorite really did trigger massive volcanic activity.

The moral of this tale is that we should not look for 'the' cause of the dinosaur extinction. It is very rare for there to be just one cause of a natural event, unlike scientific experiments which are specially set up to reveal unique explanations.

On Discworld, not only does Death come for humans, scythe in hand, but diminutive sub-Deaths come for other animals – for example the Death of Rats in *Soul Music*, from whom a single, typical quote will suffice: 'SQUEAK.'

The Death of Dinosaurs would have been something to see, with volcanoes in one hand and an asteroid in the other, trailing a cloak of ice ...

They *were* wonderfully cinematic reptiles, weren't they? Trust the wizards to get it wrong.

There is another lesson to be learned from our emphasis on the demise of the dinosaurs. Many other large and/or dramatic reptiles died out at the end of the Cretaceous, notably the plesiosaurs (famous as a possible 'explanation' of the mythical Loch Ness monster), the ichthyosaurs (enormous fish-shaped predators, reptilian whales and dolphins), the pterosaurs (strange flying forms, of which the pterodactyls appear in all the dinosaur films and are labelled, wrongly, dinosaurs), and especially the mosasaurs ...

Mosasaurs?

What were they? They were as dramatic as the dinosaurs, but they *weren't* dinosaurs. They didn't have as good a PR firm, though, because few non-specialists have heard of them. They are popularly known as fish-lizards – not as good a name as 'terrible lizard' – and it describes them well. Some were nearly as fish-like as ichthyosaurs or dolphins, some were rather crocodile-like, some were fifty-foot predators like the great white shark, some were just a couple of feet long and fed on baby ammonites and other common molluscs. They lasted a good twenty million years, and for much of that time they seem to have been the dominant marine predators. Yet most people meet the word in stories about dinosaurs, assume that the mosasaur was a not-very-interesting kind of dinosaur, and promptly forget them.

The other really strange thing about the K/T extinction – probably not a 'thing' in any meaningful sense, because in this context a thing would be an equation of unknowns, whereas what we have is a diversity of related puzzles – is which creatures *survived* it. In the sea, the ammonites all died out, as did the other shelled forms like belemnites – unrolled ammonites – but the nautilus came through, as did the cuttlefish, squids, and octopuses. Amazingly the croco-diles, which to our eyes are about as dinosaur-like as you can get without actually being one, survived the K/T event with little loss of diversity. And those little dinosaurs called 'birds' came through pretty well unscathed. (There's a story here that we need to tell, quickly. Not so long ago, the idea that birds are the living remnants of the dinosaurs was new, controversial, and therefore a hot topic. Then it rapidly turned into the prevailing wisdom. New fossil dis-coveries, however, have shown conclusively that the major families of modern birds diverged, in an evolutionary sense, long before the K/T event. So they aren't remnants of the dinosaurs that otherwise died – they got out early by ceasing to be dinosaurs at all.)

Myths, not least *Jurassic Park* itself, have suggested that dinosaurs are not 'really' extinct at all. They survive, or so semi-fact semi-fic-tion accounts lead us to believe, in *Lost World* South American valleys, on uninhabited islands, in the depths of Loch Ness, on other planets, or more mystically as DNA preserved inside blood-sucking insects trapped and encased in amber. Alas, almost certainly not. In particular, 'ancient DNA' reportedly extracted from insects fossilized in amber comes from modern contaminants, not prehistoric organisms – at least if the amber is more than a hun-dred thousand years old.

Significantly, no one has made a film bringing back dodos, moas, pygmy elephants, or mosasaurs – only dinosaurs and Hitler are popular for the reawakening myth. Both at the same time would be a good trick.

Dinosaurs are the ultimate icon for an evolutionary fact which we generally ignore, and definitely find uncomfortable to think about: *nearly all species that have ever existed are extinct*. As soon as we realize that, we are forced to look at conservation of animal species in new ways. Does it really matter that the lesser spotted pogo-bird is down to its last hundred specimens, or that a hundred species of tree-snail on a Pacific island have been eaten out of exis-tence by predators introduced by human activity? Some issues, like the importation of Nile Perch into Lake Victoria in order to improve the game fishing – which has resulted in the loss of many hundreds of fascinating 'cichlid' fish species – are regretted even by the people responsible, if only because the new lake ecosystem seems to be much less productive. Everyone (except purveyors of bizarre ancient 'medicines', their even more foolish customers, and some unreconstructed barbarians) seems to agree that the loss of magnificent creatures like the great whales, elephants, rhinos and of course plants like ginkgoes and sequoias would be a tragedy. Nevertheless, we persist in reducing the diversity of species in ecosystems all around the planet, losing many species of beetles and bacteria with hardly any regrets.

From the point of view of the majority of humans, there are 'good' species, unimportant species, and 'bad' species like smallpox and mosquitoes, which we would clearly be better off without. Unless you take an extreme view on the 'rights' of *all* living creatures to a continued existence, you find yourself having to pass judgment about *which* species should be conserved. And if you do take such an extreme view, you've got a real problem trying to preserve the rights of cheetahs *and* those of their prey, such as gazelles. On the other hand, if you take the task of passing judgment seriously, you can't just assume that, say, mosquitoes are bad and should be eliminated. Ecosystems are dynamic, and the loss of a species in one place may cause unexpected trouble elsewhere. You have to examine the unintended consequences of your methods as well as the intended ones. When worldwide efforts were made to eradicate mosquitoes, with the aim of getting rid of malaria, the preferred route was mass sprayings of the insecticide DDT. For a time this appeared to be working, but the result in the medium term was to destroy all manner of beneficial insects and other creatures, *and* to produce resistant strains of mosquitoes which if anything were worse than their predecessors. DDT is now banned worldwide – which unfortunately doesn't stop some people continuing to use it.

In the past, the environment was a context for us – we evolved to suit *it*. Now we've become a context for the environment – we change it to suit *us*. We need to learn how to do that, but going back to some imaginary golden age in which primitive humans allegedly lived in harmony with nature isn't the answer. It may not be politically correct to say so, but most primitive humans did as much environmental damage as their puny technology would allow. When humans came to the Americas from Siberia, by way of Alaska, they slaughtered their way right down to the tip of South America in a few thousand years, wiping out dozens of species – giant tree sloths and mastodons (ancient elephants, like mammoths but different), for example.

Some scientists have disagreed with this interpretation, claiming that humans could never have made that great an impact. In 2001, to test that claim, John Alroy simulated the effect of hunting in computer models of 41 North American species. It turned out that extinctions were virtually unavoidable, especially for the larger animals. Even highly incompetent hunters would have wiped them out. The simulations correctly 'postdicted' the fate of 32 of the 41 species, which adds credibility to their conclusions. 'Whodunnit?' *New Scientist* asked, offering its own suggestion: 'Mr Sapiens in the Americas with a large axe.'

There have been many similar examples of ecological devastation wrought by 'primitive' humans. The Anasazi Indians in the southern part of today's USA cut down forests to build their cliff dwellings, creating some of the most arid areas of the United States. The Maoris killed off the moas. Modern humans may be even more destructive, but there are more of us and technology can amplify our actions. Nevertheless, by the time humans were able to articulate the term 'natural environment', there wasn't one. We had changed the face of continents, in ways big and small.

To live in harmony with nature, we must know how to sing the same song as nature. To do that, we must *understand* nature. Good intentions aren't enough. Science might be – if we use it wisely.

THIRTY-NINE
BACKSLIDERS

 GLOOM HAD SETTLED OVER THE WIZARDS. Some of them had even refused a third helping at dinner.

'It's not as if they were *very* advanced,' said the Dean, in an attempt to cheer everyone up. 'They weren't even using metal. And their writing was frankly nothing but pictograms.'

'Why doesn't that sort of this thing happen here?' said the Senior Wrangler, merely toying with his trifle.

'Well, there have been historical examples of mass extinction,' said Ponder.

'Yes, but only as a result of argumentative wizardry. That's quite different. You don't expect rocks to drop out of the sky.'

'You don't expect them to stay *up*,' said Ridcully. 'In a *proper* universe, the turtle snaps up most of them and the elephants get the rest. Protects the world. Y'know, it seems to me that the most sensible thing any intelligent lifeform could do on that little world would be to get off it.'

'Nowhere to go,' said Ponder.

'Nonsense! There's a big moon. And there's other balls floating around this star.'

'All too hot, too cold, or completely without atmosphere,' said Ponder.

'People would just have to make their own entertainment. Anyway ... there's plenty of other suns, isn't there?'

'All far too far away. It would take ... well, lifetimes to get there.'

'Yes, but being extinct takes forever.'

Ponder sighed. 'You'd set out not even knowing if there's a world you could live on, sir,' he said.

'Yes, but you'd be leavin' one that you'd know you couldn't,' said Ridcully calmly. 'Not for any length of time, anyway.'

'There are new lifeforms turning up, sir. I went and checked before dinner.'

'Tell that to the lizards,' sighed the Senior Wrangler.

'Any of the new ones any good?' said Ridcully.

'They're … more fluffy, sir.'

'Doin' anything interesting?'

'Eating leaves, mainly,' said Ponder. 'There are some much more realistic trees now.'

'Billions of years of history and we've got a better tree,' sighed the Senior Wrangler.

'No, no, that's got to be a step in the right direction,' said Ridcully, thoughtfully.

'Oh? How so?'

'You can make paper out of trees.'

The wizards stared into the omniscope.

'Oh, how nice,' said the Lecturer in Recent Runes. 'Ice *again*. It's a long time since we've had a really big freeze.'

'Well, look at the universe,' said the Dean. 'It's mainly freezing cold with small patches of boiling hot. The planet's only doing what it knows.'

'You know, we're certainly learning a lot from this project,' said Ridcully. 'But it's mainly that we should be grateful we're living on a *proper* world.'

A few million years passed, as they do.

The Dean was on the beach and almost in tears. The other wizards appeared nearby and wandered over to see what the fuss was about.

Rincewind was waist deep in water, apparently struggling with a medium-sized dog.

'That's right,' the Dean shouted. 'Turn it round! Use a stick if you have to!'

'What the thunder is going on here?' said Ricully.

'Look at them!' said the Dean, beside himself with rage. 'Backsliders! Caught them trying to return to the ocean!'

Ridcully glanced at one of the creatures, which was lying in the shallows and chewing on a crab.

'Didn't catch them soon enough, did you,' he said. 'They've got webbed paws.'

'There's been too much of this sort of thing lately!' snapped the Dean. He waved his finger at one of the creatures, who watched it carefully in case it turned out to be a fish.

'What would your ancestors say, my friend, if they saw you rushing into the water just because times are a bit tough on land?' he said.

'Er ... '"Welcome back"?' suggested Rincewind, trying to avoid the snapping jaws.

''Long time no sea'?' said the Senior Wrangler, cheerfully.

The creature begged, uncertainly.

'Oh, go on, if you must,' said the Dean. 'Fish, fish, fish ... you'll turn into a fish one of these days!'

'Y'know, going back to the sea might not be a bad idea,' said Ridcully, as they strolled away along the beach. 'Beaches are edges. You always get interestin' stuff on the edge. Look at those lizards we saw on the islands. Their world was *all* edges.'

'Yes, but giving up the land to just go swimming around in the water? I don't call *that* evolution.'

'But if you go on land where you have to grow a decent brain and some cunning and a bit of muscle in order to get anything done, and then you go back to see the sea where the fish have never had to think about anything very much, you could really, er, kick butt.'

'*Do* fish have – ?'

'All right, all right. I meant, in a manner of speaking. It was just a thought, anyway.' Uncharacteristically, the Archchancellor frowned.

'Back to the sea,' he said. 'Well, you can't blame them.'

FORTY

MAMMALS
ON THE MAKE

 AFTER THE DINOSAURS CAME THE MAMMALS –

Not exactly.

Mammals constitute the most obvious class of animal alive on Earth today. When we say 'animal' in ordinary conversation, we're mostly referring to mammals – cats, dogs, elephants, cows, mice, rabbits, whatever. There are about 4,000 species of mammals, and they are astonishingly diverse in shape, size, and behaviour. The largest mammal, the blue whale, lives in the ocean and looks like a fish but isn't; it can weigh 150 tons (136,000 kg). The smallest mammals, various species of shrew, live in holes in the ground and weigh about half an ounce (15 g). Roughly in the middle come humans which, paradoxically, have specialized in being generalists. We are the most intelligent of the mammals – sometimes.

The main distinguishing feature of mammals is that when they are young their mother feeds them on milk, produced by special glands. Other features that (nearly) all mammals have in common include their ears, specifically the three tiny bones in the middle ear known as the anvil, stirrup, and hammer, which send sound to the eardrum; hair (except on adult whales); and the diaphragm, which separates the heart and lungs from the rest of the internal organs. Virtually all mammals bear live young: the exceptions are the duck-billed platypus and the echidna, which lay eggs. Another curious feature is that mammalian red blood cells lack a nucleus, whereas the red cells of all other vertebrates possess a nucleus. All is this is evidence for a lengthy common evolutionary history, subject to a few unusual events of which the most significant was the early separation of Australia from the rest of Gondwanaland, the southern half of the original supercontinent Pangaea. Modern studies of mammalian DNA confirm that basically we are all one big happy family.

When the dinosaurs died out, the mammals had a field day. Released from dinosaurian thrall, they could occupy environmental niches that, only a few million years before, would merely have presented a dinosaur with an easy meal. It seems likely that the current diversity of mammals has a lot to do with the suddenness with which they came into their kingdom – for a while, almost any lifestyle was good enough to make a living. However, it would be wrong to imagine that the mammals came into existence to fill the gaps left by the vanished dinosaurs. Mammals had coexisted with dinosaurs for at least 150 million years.

Harry Jerison has suggested that before the dinosaurs became really dominant, many mammals were able to make their living in daylight, and they evolved good eyesight to do so. As the dinosaurs became a bigger and bigger problem, the mammals adopted a lower profile, mostly staying hidden undergound during the day. If you're a nocturnal animal, you rely on a really good sense of hearing, so evolutionary pressures then equipped the mammals with excellent ears – including those three little bones. However, they retained their eyesight. So when the mammals again dared to venture out into the daylight, they had good eyesight *and* good hearing. The combination gave them a substantial advantage over most remaining competitors.

Mammals evolved from an order of Triassic reptiles known as therapsids – small, quick-moving hunters, mostly, though some were herbivores. Compared to other reptiles, the therapsids were not especially impressive, but their low-profile lifestyle led, in stages, to the distinctive features of mammals. A diaphragm leads to more efficient breathing, useful if you need to run fast. It also lets the young animals continue to breathe while mother is feeding them her milk – changes to animals 'co-evolve' as a whole suite of cooperative attributes, not one at a time. Hair keeps you warm, and the warmer you are, the faster all your bodily parts can move ... and so on.

All this makes it difficult to decide when the mammal-like reptilian ancestors of the therapsids became reptile-like mammals ... but, as we've said, humans have problems with becomings. There *was* no such point: instead, there was a mostly gradual, but occasionally bumpy, transition.* The earliest fossils that can definitely be identified as mammals come from 210 million years ago – creatures rejoicing in the name 'morganucodontids'. These were shrews, probably nocturnal, probably insect-eaters, probably egg-layers. Darwin's detractors objected to having apes as their ancestors: heaven knows what they would have thought about bug-eating egg-laying shrews. But there's good news too, if you're of that turn of mind, because morganucodontids were brainy. Not especially brainy for a shrew, but brainy compared to the reptiles from which they evolved. Admittedly, this was largely because the therapsids were as thick as two short … er, slices of tree-fern, but it was a start.

How do we know that these early shrews were true mammals? One of the bits of an animal that survives as a fossil far more often than any other bit is the tooth. This is why palaeontologists use teeth, above all else, to identify species of long-dead animals. There are plenty of species for which the sole evidence is a tooth or two. Fortunately, you can tell a lot about an animal by its teeth. On the whole, the bigger the tooth is, the bigger the animal – an elephant's tooth today is a lot bigger than an entire mouse, so whatever animal it came from, it couldn't be mouse-sized. If you can find a jawbone, a whole array of teeth, all the better. The shape of a tooth tells us a lot about what the animal ate – grinding teeth are for plants, slicing teeth are for meat. The arrangement of teeth in a jawbone tells us a lot more. The morganucodontids made a major breakthrough in tooth design: teeth that interlocked when the jaws were brought together, very effective at cutting bits off meat or insects. They also paid a heavy price for their teeth, one that we still pay today. Reptiles continually produce new teeth: as old ones wear down, they get replaced. We produce just two set of teeth: milk teeth as children and the real thing as adults. When our adult teeth wear

* OK, if you *insist* … Our favoured line here is 'hairy'. But hairs don't fossilize, so how can you tell? If you have hair, you need grooming. All over the body. This requires flexible backbones, and you can tell how flexible they are from the shape of the vertebrae. Which do fossilize. (Sometimes scientists can be *very* ingenious.) Evolution crossed *that* line about 230 million years ago.

out, the only replacements available are artificial. Blame the morganucodontids for this: if you want to take advantage of precisely
interlocking teeth, you have to maintain that precision, which is
impractical if you keep discarding teeth and growing new ones. So
they grew only two sets of teeth, and we have to do likewise.

From this we can deduce more. With only two sets of teeth, the
morganucodontids had to have some special trick for feeding their
young, something different from the reptiles with their continuous
succession of teeth. There isn't room for a full set of adult teeth in
a baby shrew, and if teeth only come in two stages, you can't add the
odd one every so often as the jaw grows bigger. The easy solution is
to have babies with no teeth at all, to start with. But what can they
then eat? Something nutritious and easily digested – milk. So we
think that milk-production evolved *before* those high-precision
interlocking teeth. This is one reason why the morganucodontids
are definitely placed among the mammals.

Amazing what you can learn from a few teeth.

As they prospered and diversified, mammals evolved into two main
types: placental mammals, where the mother carries the young in
her uterus, and marsupials, where she carries them in a pouch. The
marsupial that springs most readily to mind is the kangaroo – possibly because it springs most readily to almost anything, as for
example in *The Last Continent*:

> 'And … what's kangaroo for "You are needed for a quest of
> the utmost importance"?' said Rincewind, with guileful
> innocence.
>
> 'You know, it's funny you should ask that –'
>
> The sandals barely moved. Rincewind rose from them like
> a man leaving the starting blocks, and when he landed his feet
> were already making running movements in the air.
>
> After a while the kangaroo came alongside and accompa-
> nied him in a series of easy bounds.
>
> 'Why are you running away without even listening to what
> I have to say?'

I've had long experience of being me,' panted Rincewind.
'I *know* what's going to happen. I'm going to be dragged into
things that shouldn't concern me. And you're just a halluci-
nation caused by rich food on an empty stomach, so don't try
to stop me!'

'Stop you?' said the kangaroo. 'When you're heading in
the right direction?'

Australia alone has over a hundred species of marsupials – in fact
most native Australian mammals are marsupials. Another seventy or
so are found in the same general region – Tasmania, New Guinea,
Timor, Sulawesi, various smaller neighbouring islands. The rest are
opossums and some diminutive ratlike creatures, mainly in South
America, though ranging into Central America and for one species
of opossum right up into Canada.

It looks as though placental mammals generally win out against
marsupials, but the difference isn't so great, and if there *aren't* any
competing placental mammals then marsupials do very well indeed.
There are even some close parallels between marsupials and placentals – a good example is the koala 'bear', which isn't a true bear
but looks like an unusually cuddly one.

Most marsupials resemble 'parallel' placentals; a very curious
case is the thylacine, otherwise known as the Tasmanian tiger or
Tasmanian wolf, which is distinctly wolflike and has a striped rear.
The thylacine was officially declared extinct in 1936, but there are
persistent reports of occasional sightings, and suitable habitat still
exists, so don't be surprised if the thylacine makes a comeback.
National Park Ranger Charlie Beasley reported watching one for
two minutes in Tasmania in 1995. Similar sightings have been
reported from Queensland's Sunshine Coast since 1993: if these
sightings are genuine, they are probably of thylacines whose recent
ancestors escaped from zoos.

Why such a concentration of marsupials in Australia? The fossil
record makes it clear that marsupials originated in the Americas –
most probably North America, but that's not so certain. Placentals
arose in what is now Asia, but was then linked to the other conti-

nents, so they spread into Europe and the Americas. Before placental mammals really got going in the Americas, marsupials migrated to Australia by way of Antarctica, which in those days wasn't the frozen wasteland it is now. Australia was already moving away from South America, but hadn't yet gone all that far, and neither had Antarctica, so presumably the migration involved 'island hopping', or taking advantage of land bridges that temporarily rose from the ocean. By 65 million years ago – oddly enough, the time that the dinosaurs died out, though that's probably not significant – Australia was well separated from the other continents, Antarctica included, and Australian evolution was pretty much on its own.

In the absence of serious competition, the marsupials thrived – just as ground birds did in New Zealand, and for the same reason. But back in the Americas and elsewhere, the superior placental mammals ousted the marsupials *almost* completely.

Until a few years ago it was assumed that the placentals never made it to Australia at all – except for the *very* late arrival of rodents and bats from South East Asia about 10 million years ago, and subsequent human introduction of species like dogs and rabbits. This theory was demolished when Mike Archer found a single fossil tooth at a place called Tingamarra. The tooth is from a placental mammal, and it is 55 million years old.

From the form of the tooth it is clear that this mammal had hooves.

Did a lot of placental mammals accompany the marsupials on their migration Down Under? Or was it just a few? Either way, why did the placentals die out and the marsupials thrive?

We have no idea.

Early marsupials probably lived in trees, to judge by their forepaws. Early placentals probably lived on the ground, especially in burrows. This difference in habitat allowed them to coexist for a long time. Marsupial extinctions in the Americas were helped along by humans, who found marsupials especially easy to kill.

In Australia, about 40,000–50,000 years ago, there were sudden extinctions of *Genyornis*, the heaviest bird ever, and of the marsupial equivalent of a lion. Again, evidence suggests that humans were

THE SCIENCE OF DISCWORLD

responsible, but the theory was hotly disputed because it was difficult to date the events accurately. And, we suspect, because many
people wish to believe that all 'primitive' humans lived in exquisite
harmony with their environment. In 2001 Linda Ayliffe and
Richard Roberts used two accurate dating methods to find out
when 45 species, found as fossils at 28 separate sites, disappeared.
They *all* went extinct 46,000 years ago – just after the Aborigines,
the first humans to reach Australia, arrived.

Later arrivals were no better. When European settlers turned up,
from 1815 onwards, they very nearly wiped out numerous marsupial species.

The evolutionary history of the placental mammals is controversial
and has not been mapped out in detail. An early branch of the family tree was the sloths, anteaters, and armadillos – all animals that
look 'primitive', even though there's no earthly reason why they
should, because today's sloths, anteaters, and armadillos have
evolved just as much as today's everything else's, having survived
over the same period.

Mammals really got going during the early Tertiary period,
about 66 to 57 million years ago. The climate then was mild, with
deciduous forests at both poles. It looks as if whatever killed the
dinosaurs also changed the climate, so that in particular it was much
more rainy than it had been during dinosaur times, and the rainfall
was distributed more evenly throughout the year, instead of all
coming at once in a rainy season. Tropical forests covered much of
the planet, but they were mainly inhabited by tiny tree-dwelling
mammals. No big carnivores, not even big plant-eaters … no leopards, no deer, no elephants. It took the mammals several million
years to evolve bigger bodies. Possibly the forests were much denser
than they had been when there were dinosaurs around, because
there weren't any big animals to trample paths through them. If so,
there was less incentive for a big animal to evolve, because it wouldn't be able to move easily through the forest.

Once mammalian diversity started to get going, it exploded.
There were tigerlike animals and hippolike animals and giant

322

weasels. By modern standards, though, they were all a bit lumpish and cumbersome – nothing as graceful as the slim-boned creatures that came later, such as gazelles.

By 32 million years ago, Antarctica had reverted to being an ice-cap, and the world was cooling. Mammalian evolution had settled down, and what changes did occur were relatively small. There were bear-dogs and giraffe-rhinoceroses and pigs the size of cows, llamas and camels and sylphlike deer, and a rabbit with hooves. By 23 million years ago, the climate was warming up again. Antarctica had separated from South America, making big changes to the flow of ocean currents: now cold water could go round and round the south pole indefinitely. The sea level fell as water got locked up in ice at the poles; with more land exposed and less ocean the climate became more extreme, because land temperatures can change more quickly than sea ones. Falling sea levels opened up land bridges between previously isolated continents; isolated ecologies started to mix up as animals migrated along the new connections. And round about this time, the evolution of some mammals took an unusual turn. A U-turn.

They went back to the sea.

The land animals had originally come out of the sea – despite the wizards' best efforts to stop them. Now a few mammals decided they'd be better off going back there. The wizards consider such a tactic to be a spineless piece of backsliding, giving up and going back home. Even to us it looks like a retrograde step, almost counter-evolutionary: if it was such a good idea to come out of the oceans in the first place, how could it be worthwhile to go back again? But the evolutionary game is played against a changing background, and the oceans had changed. In particular, the available *food* had changed. So in the mid-Eocene we find the earliest fossils of whales, such as the sixty-five (20 m) long *Basilosaurus*, which had a pair of tiny legs at the base of its long tail. We've found fossils of its ancestors, and they really did look like small dogs.

The Mediterranean sea was dammed, Africa came into contact with Europe, and creatures previously confined to Africa spread into Europe, among them elephants – and apes. Horses evolved, as did true cats (such as the famous sabre-toothed tiger). By five million years ago, most of today's mammals were represented in recognizable form, and the climate had become similar to today's.

The scene was set for the evolution of humans.

Not that it had all been set up *in order* to lead to us, you appreciate. Our early ancestors just happened to be in a position to take advantage of the world as it then was. They did so.

We can trace the ancestry of modern mammals – indeed all living creatures that still exist today – by mapping out changes in their DNA. The rate at which DNA *mutates* – acquires random errors in its code – leads to a 'DNA clock' that can be used to estimate the timing of past events. When this technique was first discovered, it was widely hailed as a precise and therefore uncontroversial way to resolve difficult questions about which animals' ancestors were more closely related to what. It is now becoming clear that precision alone cannot provide definitive answers to such questions.

The issue of interpretation – what does this result *mean*? – can still be controversial, even if the result itself can be made precise. For example, S. Blair Hedges and Sudhir Kumar have applied the DNA clock to 658 genes in 207 species of modern vertebrates: rhinos, elephants, rabbits, and so on. Their results suggest that many of these lineages were around at least 100 million years ago, coexisting with the dinosaurs – though no doubt the early elephant and rhino ancestors were rather small. The fossil record agrees that there were mammals then – but not those. The molecular biologists claim that the fossil record must be misleading; palaeontologists are convinced that the DNA clock sometimes ticks faster and sometimes ticks slower. The debate continues – but for what it's worth, our money is on the palaeontologists.

One big surprise about mammal DNA is how much of it there is. You might expect a sophisticated creature like a mammal to be 'hard to build' and therefore require more DNA, just as the blueprint for a jumbo jet has to be more complicated than that for a kite.

Not so.

Mammals have *less* DNA – shorter genomes – than many apparently simpler animals, for example frogs and newts.

There's a good reason for this apparent paradox, and it illuminates the difference between DNA and a blueprint. DNA is more like a recipe – and a recipe that makes a lot of assumptions about what else you have in your kitchen, so that none of that needs to be spelled out in the recipe book. In essence, the kitchen for mammals has a really well controlled oven, capable of ensuring nice, even cooking temperature, so a whole lot of tricks about what to do if the temperature changes need not be mentioned.* In the frog kitchen, on the other hand, the temperature goes up and down depending on the time of day and the weather, so the recipe has to deal with all contingencies, requiring more DNA code. By 'kitchen' here we mean the environment in which the embryonic animal has to develop. For a frog, the kitchen is a pond. For a mammal, the kitchen is mother.

Mammals evolved good temperature control – unlike the reptiles, they are warm-blooded, but what matters is not so much being *warm*, as being controllable. Frog DNA is full of genes for making lots of different enzymes, together with instructions along the lines of 'use enzyme A if the temperature is lower than 6°C, use B if the temperature is between 7°C and 11°C, use C if the temperature is between 12°C and 15°C ...' Mammal DNA just says 'Use enzyme X', knowing that mother will take care of temperature variations. Frog DNA is a rocket: mammal DNA is a space elevator.

How did this change take place? Perhaps when mammals first evolved, their DNA gained extra instructions, but after temperature control evolved, a lot of the DNA became redundant, and it either got dumped or got subverted to other uses. On the other hand, we have no idea what the DNA of early mammals actually looked like – maybe it was *all* shorter in those days, maybe today's frogs and newts have much more extensive recipes than ancient ones. But on balance it seems more likely that mammals just eliminated a lot of surplus instructions

---

* How many recipe books do you have that tell you to boil water, but *never* specify the altitude at which this should be done? It matters: higher up, water boils at lower temperatures.

Modern technology uses the same trick. Because the machinery that makes today's consumer goods is extremely precise and accurate, those goods can be *simpler* than they were in the past. A soft drinks can, for example, is little more than a piece of aluminium that has been formed into a cylinder, with another flat bit on top to act as a lid, a weak line for the tab to tear along, and a ring (or nowadays a lever) attached to the tab. It replaces the bottle, which consisted of two or more bits of moulded glass 'welded' together, a metal cap, and a slice of cork. The simplicity of the can comes at a price: *very* careful control of the forming process.

There are many scientists who insist that an organism's DNA determines everything about it – even though it manifestly does not – and they argue that the mother's temperature-control system is included in *her* DNA recipe. This may well be true, but even if it is, 'this organism's' DNA has somehow migrated to another organism (mother, not her offspring). As soon as two generations are involved in implementing the genetic blueprint, a gap opens up into which things can be inserted that are not genetic at all. We've already mentioned several, for example prions in the reproduction of yeast.

Our mammalian ancestry may even be responsible for one of the more bizarre modern myths, persistent tales of people being abducted by aliens. Ufologists allege that one American in twenty now claims to have undergone such an experience (but they would, wouldn't they?). If true, this figure would be a remarkable and not very happy comment either on the critical faculties of that great nation or on the habits of an unknown spacefaring species.

As it happens, the figure is bogus. It originates in a Roper poll of 1994, which revealed that one American in fifty had undergone such an experience. But, as Joel Best pointed out in his book *Damned Lies and Statistics* in 2001, the number of people who claimed to have been abducted by aliens was actually *zero*. The pollsters, worried that a direct question about aliens would put people off, used five 'symptoms' of abduction instead. Anyone who scored sufficiently highly on those symptoms was *deemed* to have undergone an abduction experience.

The questions were things like 'Have you ever woken up paralysed with the sense of some strange presence in the room?' This sensation is typical of 'sleep paralysis', the most obvious rational explanation of abduction experiences, which we describe shortly. So really the Roper poll was a survey about sleep paralysis. Only the researchers thought that it had anything to do with alien abduction. The subjects had more sense.

Be that as it may, a lot of people are convinced that strange aliens, usually with big black eyes and pear-shaped heads like the ones in *Close Encounters of the Third Kind*, landed a UFO near them, loaded them on board, and took them for a flight round the solar system while carrying out weird experiments, often of a sexual nature, on them. After which they were calmly returned to the very spot from which they had been abducted, as if absolutely nothing had happened.

The first thing to say is that without doubt many of these experiences are false. Ian once did a radio broadcast which included a woman who had undergone a convincing experience of being abducted – except that she knew she hadn't really been, because her family told her she'd been asleep beside the fire the whole time. Jack once met a woman who claimed that the aliens abducted her and took away her baby. So he asked a question that nobody else had thought to ask, the woman included: 'Were you pregnant?'

'No.'

The point is that to the victims, the experience *felt* real. Even though logic told them it couldn't have happened, they either didn't apply the logic, or they did but still remembered the experience vividly. We deduce that the human mind sometimes has vivid memories that do not correspond to real events. Of course we must also observe that just because some alien abductions aren't real, that doesn't imply they all aren't. However, if we can find a sensible mechanism for otherwise reasonable people *believing* that they really were carted off in a UFO, then the burden of proof shifts dramatically and evidence of abduction stronger than sincere expressions of belief becomes necessary.

Reports of alien abductions are not new. Back in the Middle

Ages, however, they would have been either flights on witches' broomsticks or encounters with fabulous creatures like the succubus, a demon in a woman's body who allegedly had sex with men while they slept. The witches of Discworld employ broomsticks for transport only. The sex bit doesn't appeal to them at all – except for Nanny Ogg, of course.

Folk tales of succubi and their like can be found worldwide. In Newfoundland people tell of an ancient hag sitting on their chests at night, and in Vietnam they speak of the 'grey ghost'. What seems to be going on is some common mental pattern, overlaid with cultural influences. That's why abductions by witches riding broomsticks have gone out of vogue, but abductions by aliens riding UFOs are flavour-of-the-decade.

Susan Blackmore thinks that all of these experiences are, and were, caused by sleep paralysis. This is a feature of the mind that prevents sleeping people from moving their limbs as they would if they were acting out their dreams. Such a 'mental switch' is important for any animal that dreams: you don't really want to go sleepwalking out of your cosy burrow and straight down a predator's throat. Plenty of mammals dream – most of us have seen a cat or dog asleep with its legs twitching, and the evidence from recordings of the brain's electrical activity is that the animals are engaged in something that closely resembles the brain activity of a dreaming human. We can't be sure whether cats have visual dreams like we do, but sleep and dreaming take place in primitive parts of the brain, so they probably go back a long way in our evolutionary history. At any rate, if the sleep paralysis system malfunctions, people who are partially awake may undergo sleep paralysis. Experiments show that in such cases they typically get a strong impression that 'somebody is there'.

This feature of the human mind may go back to the time, just after the meteorite hit, when the nocturnal mammals suddenly awoke in a world without dinosaurs. Their senses of hearing and sight, previously separate from each other because they had evolved at very different periods and in very different circumstances, would have become linked together. When their ears heard something strange, their visual sense would kick in and make them feel that they could *see* what was causing it. We inherited this tendency, but we interpret it in terms of the current culture: bogeymen, witches, maybe even dragons a few centuries ago, aliens with big black eyes today. The sexual link is straightforward, too: dreams about sex are very common anyway.

Oh, yes, one more thing: since we've all watched *Close Encounters*, we know exactly what an alien must look like … just as everyone used to know that witches soared through the air on a broomstick. So our visual system knows what shape it should give to whatever it *sees* when we get that funny feeling that something is haunting us. And flying saucers have come on nicely, too, from being the rivet-studded things that were all the rage in galactic circles in the early Fifties.

Stories of people seeing ghosts may well have the same explanation. You've read the tales, you know what a ghost ought to look like (maybe you watched *Ghostbusters* or a Stephen King movie), and you're trying to sit up all night in the Haunted House. You're thinking about ghosts, about headless horsemen and Elizabethan ladies who walk through walls and go transparent – and then you start to doze off because it's 2 am and you've been up all night … The sleep paralysis circuit glitches … *Aaaaagh!*

FORTY-ONE

DON'T
PLAY GOD

THE ARCHCHANCELLOR WAS RATHER QUIET OVER TEA.

Eventually he said, 'Can we *stop* this project, Ponder?'

'Er ... are you *sure*, sir?'

'Well, what is it achieving? I mean, *really*? Y'know, I thought, all you had to do is get a world working, and before you could say "creation" there'd be some creature who'd stand up, getting a grip on its surroundings, gaze with a certain amount of intelligence and awe at the infinite sky and say –'

'– that thing's getting bigger, I wonder if it's going to hit us,' said Rincewind.

'Rincewind, that remark was extremely cynical and accurate.'

'Sorry, Archchancellor.'

Ponder's lips were moving quietly as he worked things out.

'We *could* start running it down, yes. The thaumic reactor hasn't been putting so much into it in the last week. We've nearly used up the fuel.'

'Really?'

'The squash court will have a rather high thaumic index, sir, so whoever goes in to pull the switch will suffer a certain amount of –'

There was the sound of something spinning. The wizards looked at Rincewind's chair, which finally fell over on to the flagstones. Of its former occupant there was no sign, although there was the distant sound of a slamming door.

The Dean sniffed.

'Strange behaviour,' he said.

'I suggest we give it one more day of our time,' said Ridcully. 'I was hoping we might create a world, gentlemen, but instead it's clear to me that any life in this universe has to get used to living in ... in some kind of huge celestial snow globe. Fire and ice, ice and fire. Gentlemen, round worlds are intrinsically flawed. If there's any hidden gods on ours, they're pretty damn well hidden.'

'The Omnians say "Don't play God. He always wins",' said the Senior Wrangler.

'I dare say,' said Ridcully. 'So ... one more day, gentlemen? And then we can get on with something sensible.'

The red sun rose quickly over the parched veldt. The apes stirred in their cave, which was little more than a rocky overhang, and saw the big black rectangle looming over them.

The Dean tapped it with his pointer.

'*Do* try to pay attention today, will you?' He turned and chalked rapidly across the blackboard. 'Here we have R ... O ... C ... K, rock. Can anyone tell me what you do with it? Anyone? Anyone? Look, stop doing that, will you?' He tried to hit an ape with his virtual pointer, and then flung it away in disgust. It vanished.

'Filthy little devils,' he muttered.

'Not getting anywhere, Dean?' said Ridcully, appearing beside him.

'*No*, Archchancellor. I've *tried* to explain to them that they've probably got just a few million years, and that's pretty hard to do in sign language, let me tell you. But the only word they know is S-E-X, and they don't waste time *spelling* it, oh no! For this I skipped breakfast?'

'Never mind. Let's see how the Senior Wrangler's getting on.'

'They're just bad copies of humans, if you ask me –'

The wizards vanished.

One of the apes knuckled over to the blackboard, and watched it disappear from view as HEX completed the spell.

He hadn't the faintest idea what had been happening, but he *had* been impressed by the stick that had been waved about. That seemed to have gone now. That didn't worry the ape, which knew about things vanishing – often, these days, a member of the clan would vanish overnight, with a lot of snarling in the shadows.

There was probably something you could do with a stick, he thought. Hopefully, it might involve sex.

He poked around in the debris and found not a stick but a dried-up thighbone, which had a sufficiently stick-like shape.

He rattled it on the ground a few times. It didn't do anything much. Then he reluctantly decided it would probably be impossible to mate with at the moment, and hurled it high into the air.

It rose, turning over and over.

When it fell, it knocked him unconscious.

The Senior Wrangler was sitting under a virtually-there beach umbrella when the other wizards arrived. He looked as downcast as the Dean.

A group of apes was playing in the surf.

'Worse than the lizards,' he said. '*They* had some style, at least. When this lot pick up anything, they try to see if they can eat it. What's the point of that?'

'Well, I suppose they can find out if it's edible,' said Ridcully.

'But they just *mess about*,' said the Senior Wrangler. 'Oh, no … here we go again …'

There was a raucous shrieking as the tribe rushed out of the waves and swung up into the nearby mangrove trees. A shadow sped beyond the surf and headed back into the blue water, to an unregarded chorus of simian catcalls and mangrove seeds.

'Oh yes, and they like throwing things,' said the Senior Wrangler.

'Seafood is good for the brain, my granny always said,' said Ridcully.

'This lot couldn't eat too much of it, then. Yell, throw things, and prod stuff to see what it does, that's the extent of their capabilities. Oh, *why* didn't we discover the lizards earlier? *They* had *class*.'

'Wouldn't have stopped the snowball,' said Ridcully.

'No. You were right, Archchancellor. It's so *pointless*.'

The three wizards stood looking gloomily out to sea. In the middle distance, dolphins stitched their way across the water.

'Should be coming up to coffee time,' said the Dean, to break the silence.

'Good thinking, that man.'

Rincewind was wandering in the next bay, staring at the cliffs. Oh, things were killed off on the Discworld, but ... well ... *sensibly*. There were floods, fires and, of course, heroes. There was nothing like a hero for a species whose number was up. But at least some actual thought went into it.

The cliff was a series of horizontal lines. They represented ancient surfaces, some of which Rincewind had virtually walked on. And in many of them were the bones of ancient creatures, turned into stone by a process Rincewind did not understand and rather distrusted. Life had some how come out of the rocks of this world, and here you could see it going back. There were whole layers of rock made out of life, millions of years of little skeletons. Faced with a natural wonder on that scale, you could only be overawed by the sheer chasms of time or else try to find someone to complain to.

A few rocks fell out, halfway up the cliff. A couple of small legs waved uncertainly in the strata, and then the Luggage tumbled out, slid down the pile of debris at the foot of the cliff, and landed on its lid.

Rincewind watched it struggle for a while, sighed, and pushed it the right way up. At least some things didn't change.

FORTY-TWO

ΛNTHILL INSIDE

 YOU KNOW WHAT'S GOING TO HAPPEN TO THE APES — they're going to turn into *us*.* But why do we have them playing in the surf? Because it's fun? Yes ... but more significantly, because the seashore is central to one of the two main theories about how our ape ancestors acquired big brains. The other, more orthodox theory places the evolution of the big brain out on the African savannahs, and we *know* that some of our ancestors lived on the savannahs because we've found fossils. Unfortunately, seashores aren't a good place to leave fossils. You often *find* them there, but that's because they were deposited when the area wasn't a seashore at all, and the sea has subsequently eroded the rocks to expose the fossils. In the absence of direct evidence of this kind, the surfing apes theory has to take second place ... but it does explain our brains rather neatly, whereas the savannah theory rather sidesteps this issue.

Our closest living relatives are two species of chimpanzee: the standard boisterous 'zoo' chimp *Pan troglodytes* and its more slender cousin the bonobo (or pygmy) chimp *Pan paniscus*. Bonobos live in very inaccessible parts of Zaire, and weren't recognized as a separate species of chimpanzee until 1929. We can to some extent unravel the past evolutionary history of the great apes by comparing their DNA sequences. Human DNA differs from the DNA of either chimpanzee by a mere 1.6% – that is, we have 98.4% of our DNA sequences in common with theirs. (It is interesting to speculate on what the Victorians would have made of this.) The two species of chimpanzee have DNA that differs by only 0.7%.

* To find out how, see *The Science of Discworld II: The Globe.*

Gorillas differ from us, and from both chimps, by 2.3%. For orangutans, the difference from us is 3.6%.

These differences may seem small, but you can pack an awful lot into a small percentage of an ape genome. A big chunk of what we have in common must surely consist of 'subroutines' that organize basic features of vertebrate and mammalian architecture, tell us how to be an ape, and tell us how to deal with things we've all got – like hair, fingers, internal organs, blood ... The mistake is to imagine that everything that makes us human and not a chimpanzee must live in that other 1.6% of 'special' DNA – but DNA doesn't work that way. For example, some of the genes in that 1.6% of the genome may organize the other 98.4% in a completely new way. If you look at the computer code for a wordprocessor and a spreadsheet, you'll find they have an awful lot in common – routines for reading the keyboard, printing to the screen, searching for a given text string, changing fonts to italic, responding to a click on the mouse ... but this doesn't mean that the *only* distinction between a spreadsheet and a wordprocessor lies in the relatively few routines that are different.

Since evolution involves changes to DNA, we can use the sizes of those differences to estimate when various ape species diverged from each other. This method was introduced by Charles Sibley and Jon Ahlquist in 1973, and while it needs to be interpreted with caution, it works well here.

A convenient unit of time for such discussions is the 'Grandfather', which we define to be 50 years. It's a good human length, being about the age difference between the child and the grandparent who says 'When *I* was young ...' and passes on a sense of history. In these terms, Christ lived 40 Grandfathers ago, and the Babylonians go back about 100 Grandfathers. That's not a lot of grandads, passing down through recorded human history recollections like '... we never had any of this modern cuneiform when *I* was a lad ...' and '... bronze was good enough for me'. Human time is not very deep. We've just been good at packing a lot into it.

DNA studies indicate that the two chimp species diverged about 60,000 Grandfathers ago – three million years. Humans and chimps diverged 80,000 Grandfathers earlier – so a chain of only 140,000 grandfathers unites you and your chimplike ancestor. Who was also, we hasten to point out, a modern chimpanzee's manlike ancestor. Humans and gorillas diverged 200,000 Grandfathers ago; humans and orangutans diverged 300,000 Grandfathers ago. So among these animals, we are most closely related to a chimpanzee, and least closely related to the orangutan. This conclusion is borne out by physical appearance and habits, too. Bonobos really *like* sex.

If those times seem rather short for all the necessary evolutionary changes, bear two things in mind. First, that they were estimated by using a *realistic* rate for DNA mutations; second, that according to Nilsson and Pelger an entire *eye* can evolve in a mere 8,000 Grandfathers – and lots of different changes can, should, and *did* evolve in parallel.

The most striking feature of humans is the size of our brains: bigger, in comparison to body weight, than any other animal. Strikingly bigger. A detailed story of what makes us human must be extraordinarily complicated, but it's clear that big, powerful brains were the main invention that made it all possible. So we now have two obvious questions to think about: 'Why did we evolve big brains?' and 'How did we evolve big brains?'

The standard theory addresses the 'why'. It maintains that we evolved out on the savannahs, surrounded by lots of big predators – lions, leopards, hyenas – and without much cover. We had to become smart in order to survive. Rincewind would instantly see one flaw in this theory: 'If we were so smart, why did we *stay* on the savannahs, surrounded by lots of big predators?' But, as we've said, it fits the fossil evidence. The unorthodox theory addresses the 'how'. Big brains need lots of brain cells, and brain cells need lots of chemicals known as 'essential fatty acids'. We have to get these from our food – we can't build them ourselves from anything simpler – and they're in short supply out on the savannahs. However, as Michael Crawford and David Marsh pointed out in 1991, they are abundant in seafood.

Nine years earlier Elaine Morgan had developed Alister Hardy's theory of the 'aquatic ape': we evolved not on the savannahs, but on the seashore. The theory fits a number of human peculiarities: we like water (newborn babies can swim), we have a funny pattern of hair on our bodies, and we walk upright. Go to any Mediterranean resort and you see at once that an awful lot of naked apes think that the seashore is *the* place to hang out.

Whether the Aquatic Ape story of human origins will displace the 'savannah' theory remains to be seen, but the savannah story is in trouble from a very different direction. Phillip Tobias has challenged not the fossil record, but its interpretation. He asked a question so simple that it seems to have eluded most other workers in the field. Yes, many areas where fossils of apelike human ancestors have been found are savannahs *now*. But were they savannahs *then*? When our distant great-great-... grandparents got themselves fossilised, 2.7 million years ago, could the vegetation have been different from what it is today?

Given that the whole point is that the animals definitely were different – our ancestors, not us – it is a little surprising that this question seems not to have occurred to anyone earlier. Unfortunately, science is often like that. People specialise. Experts on prehistoric apes may not be very interested in botany.

It turns out that Sterkfontein, one of the places where fossil apes supporting the savannah theory are found, wasn't a savannah back then at all. Fossil pollen suggested it was a wooded area, and fossilised lianas clinched it. Other areas of South Africa and Ethiopia (where the famous 'Lucy' was found) show that these areas were forests when the apes lived there. The 'killer ape of the savannahs,' says Tobias, is nonsense.

And there may be some new evidence in favour of watery origins for humanity, though not necessarily the full-blooded Aquatic Ape. A common feature of fossil hominid sites is that they are all near water. This makes sense, because *Homo sapiens* needs to drink a lot, and sweats and urinates a lot. If we had evolved on savannahs, we would have annoyed the hell out of all the other animals there with our incessant peeing. And it looks like we were excellent swimmers at least a million years ago. There is evidence of human migrations to islands such as Flores, separated from Bali by a deep underwater valley. Even allowing for the lower sea-levels in the past, the new arrivals must have swum or rafted or otherwise made their way across at least 20 miles of open water.

We may not have been the Aquatic Ape, but we surely were the Damp Forest Ape. Just as bonobos, one of our two closest living relatives, are today.

Brains are fascinating. They are the physical vehicle for minds, which are even more interesting. Minds are (or, at least, give their owners the vivid impression that they are) conscious, and they have (or, at least, give their owners the vivid impression that they have) free will. Minds operate in a world of 'qualia' – vivid sense impressions like *red, hot, sexy*. Qualia aren't abstractions: they are 'feelings'. We all know what it's like to experience them. Science has no idea what makes them the way they are.

Brains, though … we can make progress on brains. On one level, brains are a kind of computational device. Their most obvious physical components are nerve cells, arranged in complicated networks. Mathematicians have studied such networks, and they find that what networks do is to carry out interesting processes. Give them an input and they will produce an output. Allow their interconnections to evolve by selecting for specific associations of input and output – such as responding to an image of a banana but not to an image of a dead rat – and pretty soon you've got a really effective banana-detector.

What makes the human brain unique, as far as we can tell, is that it has become recursive. As well as detecting a banana, it can think *about* detecting a banana. It can think thoughts about its own thought processes. It is a pattern-recognition device that has turned its attention to its own patterns. This ability is what lies behind human intelligence. It probably underpins consciousness, too: one of the patterns that the pattern-recognition device has learned to recognize is *itself*. It has become 'self-aware'.

As a result, brains operate on at least two levels. On a reductionist level they are networks of nerve cells sending each other incredibly complex but ultimately meaningless messages – like ants scurrying around inside an anthill. On another level, they are an integrated self – the anthill as a personality in its own right. Douglas Hofstadter's *Gödel, Escher, Bach* includes a sequence where Aunt Hillary (who is an anthill – use the American pronunciation of 'Aunt') has a meeting with Dr Anteater. When Dr Anteater arrives, the ants go into a panic – they change their actions. To Aunt Hillary, who operates on the emergent level, this change represents the *knowledge* that Dr Anteater has arrived. She is entirely happy to watch Dr Anteater consuming a meal of 'her' ants. Ants are a virtually inexhaustible resource – she can always breed new ones to take the place of the ones that got eaten.

The link between the ants and Hillary's 'anthilligence' is emergent – felicitously, it operates across what we have termed 'Ant Country'. The same action means one thing for the ants, but something quite different, and *transcendent*, for Hillary. Replace Hillary by yourself – your *self*, the 'you' that you feel is experiencing your thoughts – and ants by brain cells, and you're contemplating the connection between mind and brain.

Now *you've* gone self-referential.

Neural networks are what the brain is built from, but there's more to evolving a brain than just assembling big neural nets. Brains operate in terms of high-level 'modules' – a module for running, another for recognizing danger, another for putting the whole animal on the alert, and so on. Each such module is an emergent feature of a complex neural network, and it wasn't designed: it evolved. Millions of years of evolution *trained* those modules to respond instantly and exquisitely.

The modules aren't separate. They share nerve cells, they overlap, they're not necessarily a well-defined *region* in the brain – any more than 'Vodafone' is a well-defined *region* of the telephone network. According to Daniel Dennett, they are like a collection of demons, operating by 'pandemonium'. They all shout, and at any given instant, whoever shouts loudest wins (quite a lot of the Internet has borrowed this design).

Modern humanity has built a culture around those modules – an idea that we'll explore later – and in so doing has subverted them to new purposes. The module for spotting lions has become, in part, a module for reading Discworld books. The module for sensing bodily movement has, in part, turned into one for doing certain kinds of mathematics – those parts of mechanics where a physical 'feel' for the problem may well be precisely that. Our culture has rebuilt our minds, and our minds have in turn rebuilt our culture, over and over again, in each generation.

Such a radical restructuring must have simpler precursors. A key step towards the human mind was the invention of the nest. Before there were nests, baby organisms could carry out only very limited experiments in behaviour. If every time you try out a new game you get gobbled up by a python, novelty will not carry a premium. In the comfort and relative safety of the nest, however, the *error* part of trial-and-error is no longer automatically fatal. Nests let you play, and play lets you explore the phase space of possible behaviours and find new, sometimes useful, strategies. Further along the same path lies the family, the pack, and the tribe, with certain shared behaviours and mutual protection. Meerkats, a kind of mongoose, have an intricate tribal structure, and take turns doing the dangerous (because more exposed) job of Lookout.

Humans have turned such tactics into a global strategy: adults devote huge amounts of time, energy, food, and money to the task of bringing up their children. Intelligence is both a consequence of this brilliantly successful strategy, and a cause.

The Dean would be well advised to take this link between family life and intelligence into account. He's trying to educate the apes by the direct route (R ... O ... C ... K ...) but all they have on their tiny minds is S-E-X. Many school teachers will sympathize ... but if only he realized that sexual bonding is a major factor in humanoid family life, and family life engenders intelligence ...

Bonobos are the perfect model for the Dean's sex-mad apes. They are promiscuous in the extreme, making use of sex where we would be content with a smile and a wave or a gentlemanly handshake. Female bonobos have serial sex with dozens of males, or with females, almost in passing; the males do likewise. Adults engage in sexual activities with children, too. It all seems very casual. It helps bond the tribe. For them it seems to work fine.

Ordinary chimps are promiscuous by the standards of orthodox human morality, though probably no more than many humans are. Pairs of males and females will disappear together for a few days, and then form new partnerships ... Humans generally mate for life (a term meaning 'until we get fed up') and one reason is the enormous amount of effort that a human couple must put into raising the kids. Sex helps to cement the parental relationship, encouraging each parent to trust the other. This may be why, even in an allegedly sexually relaxed age, most people see extramarital flings as a form of betrayal – and why, despite that, the erring partner is more often than not allowed back into the family fold.

It's not surprising that we have sex on the brain: our brains have been moulded by sex. The Dean should let sex take its course, for intelligence will surely follow ... You just have to think on the scale of Deep Time. There's no rush.

FORTY-THREE

OOK: A SPACE ODYSSEY

 RINCEWIND SAT IN A CORNER of the High Energy Magic building. It was deserted at the moment. News had got around that the project was really being ended this time, and wizards had drifted away to lunch.

The round world spun in its protective globe and also, by means of a physics only a wizard might understand, in a space that was infinite only on the inside.

'Poor old bloody place,' he said, to the world in general. 'Never really stood a chance, did you.'

'Ook.'

It was a small grunt, from the other side of the huge room. Rincewind wandered over, and found the Librarian peering into the omniscope.

'Oh, they've got sticks now,' said Rincewind, looking down at a ragged party of apes. 'And a lot of good it'll do them, too.'

'Ook?'

'The lizards had sharp shells on the end of theirs, and are they around today? I don't think so. And the crabs were doing well. Even the blobs were trying to make a go of things. There were some bear sort of things that looked promising. Doesn't matter. One winter the snow doesn't melt, next thing there's a two-mile wall of ice laminating you to the bedrock. Or there's a funny light in the sky and then you're trying to breathe burning water.' He shook his head wearily. 'Nice place, though. Nice colours. Particularly good horizons, once you get used to them. Lots of dullness, punctuated by short periods of death.'

'Ook?' said the Librarian.

'Well, maybe they do look a bit like you,' said Rincewind. 'Most of the lizards looked a bit like the Bursar. Maybe it's just coincidence. Everything has to look like something, after all. As above, so below.'

In the omniscope, some distance behind the ape clan, something lean and powerful was tracking them in the long grass.

'Eeek!'

The Librarian thumped on the desk.

'Sorry. It's not up to me. "Live and let live", you know that's always been *my* motto. Well, "let *me* live", really, but that's almost the same thing.'

Hands waving wildly over his head, which only happened when he was really in a hurry, the Librarian ran out of the room.

Rincewind caught him up as he entered the main building, and then trotted along after him as the ape wound his way through the university's less salubrious regions, the realm of broom cupboards, old storerooms and the studies of the very much lesser members of staff. Even using all the shortcuts, it still took quite a while to reach the office of the Egregious Professor of Cruel and Unusual Geography, with the name 'Rincewind' written on it in chalk.

The orangutan flung the door back and knuckled purposefully towards the big stack of boxes.

'Er ... that's the rock collection,' said Rincewind. 'Er ... I was filing them ... er ... they belong to the University, I really don't think you should be throwing them out like that –'

'Ook!'

The Librarian straightened up, bearing aloft a couple of large rocks that Rincewind recognized as noduley, sharp, brittle, unfriendly rocks.

'Er ... why are you ...' Rincewind began.

The Librarian walked across to the Luggage and gave it a kick. The lid opened obediently, and the rocks were thrown inside. The ape went back for more flints.

'Er ...' said Rincewind, but left it at that. This did not seem to be a time to raise objections.

He had to run after the Librarian and the Luggage all the way back to the High Energy Magic Building. By the time he got there, the ape was pounding heavily on one of Hex's keyboards.

Rincewind tried again.

'Er ... should you be ...'

He was interrupted by the rattle of the machine's writing device. It spelled out: +++ New Suit Parameters Accepted +++

On the far side of the room, where the skeletal virtually-there suits flicked on the verge of non-existence, one changed shape. The shoulders widened. The arms grew longer. The legs shortened –

+++ Adjustment Complete. On You It Looks Good +++

Rincewind backed away as the Librarian, cradling a large flint nodule in each arm, stepped in the magic circle and began to shimmer as the suit enclosed him. The new parameters definitely made it look more *solid* ...

'You're not going to *interfere*, are you?' said Rincewind.

'Ook?'

'No, no, that's fine, fine, no problem at all,' said Rincewind. It is never wise to argue with an ape holding a rock. 'It's about time someone did.'

The Librarian flickered, and became a ghost in the air.

Rincewind stood alone in the empty room, whistling nervously. In its alcove, HEX began to sparkle, as it always did when it was trying to allow a wizard to interact with the project.

'Blast!' said Rincewind at last, striding over to the suits. 'He's *bound* to muck it up ...'

Lightning fried the evening sky, turning it purple and pink.

Above the little hollow in the cliff, where the tribe clustered and flinched, a sleek black shadow moved like an extension of the night. It wasn't hurrying. Dinner wasn't going anywhere. When the lightning faded its eyes gleamed for a while.

Something grabbed its tail. It spun around, snarling, and a fist extended on the end of a very long arm hit it right between the eyes, lifting it off the ledge.

It landed heavily on the ground, jerked for a moment, and lay still.

The ape horde scattered around the rocks, screaming, and then stopped to look back.

The big cat didn't move.

Another bolt of lighting hit the ground nearby, and a dead tree exploded into flame. Against the violet corona of the storm, red in the light of the burning tree, a huge figure stood holding a large stone in the crook of each arm.

As Rincewind said, it was a vision you were unlikely to forget.

Rincewind couldn't eat here. Well, not in the usual, definitive way. He thought he could probably manipulate lumps of food into his mouth, but since the food would technically remain in a different universe to his, he was afraid it might drop straight through him, to general embarrassment and the puzzlement of spectators.

Besides, he didn't feel like flame-grilled leopard.

The Librarian had been working furiously. He'd turned the area into a boot camp for people who were barely upright and wouldn't know what to do with a boot anyway. The apemen had taken to fire quite quickly, after a few misdirected attempts to eat it or have sex with it, and several of them had progressed to setting fire to themselves.

They'd learned cookery, too, initially on one another.

Rincewind sighed. He'd seen species come, and he'd seen them go, and this one could only have been put on the world for entertainment value. They had the same approach to life as clowns, with the same touch of cheerful viciousness.

The Librarian had progressed to lessons in flint-knapping, using the flints brought in via the Luggage. They'd certainly picked up the idea of hitting rocks against other rocks, or anything else in range. Sharp edges intrigued them.

Finally Rincewind wandered over to the Librarian and tapped him on the shoulder.

'We've been here all day,' he said. 'We'd better get back.'

The orangutan nodded, and stood up. 'Ook.'

'You think it'll work?'

'Ook!'

Rincewind looked back at the apemen. One of them was industriously hacking at the corpse of the cat again.

'Really? But they're just like ... hairy parrots.'

'Eek ook.'

'Well ... yes. That's true.' Rincewind took a final look at the horde. Two of them were squabbling over the meat. Monkey see, monkey do ...

'I'm glad it was *you* who said that,' he said.

Less than a Discworld second had passed by the time they returned. By the time they looked in the omniscope, several fires were already visible on the night side of the world.

The Librarian looked pleased. 'Oook,' he said.

Progress means smoke. But Rincewind was not entirely convinced. Most of the fires were forests.

FORTY-FOUR

EXTEL OUTSIDE

 PROGRESS MEANS SMOKE ... The human race has certainly made a lot of progress over the years, then. How did we do that? Because we're *intelligent*, we've got *brains*. Minds, even. But other creatures are intelligent – dolphins, especially. And all they seem to do is enjoy themselves in the sea. What have we got that they haven't?

Many discussions of the mind treat it essentially as a question about the architecture of the brain. The viewpoint is that this determines what brains can do, and then the various things that we associate with minds – the difficult problems of free will, consciousness and intelligence – come out of neurophysiology. That's one approach. The other common one is to view the problem through the eyes of a social scientist or an anthropologist. From this viewpoint the mind's capabilities are pretty much taken as 'given', and the main questions are how human culture builds on those capabilities to create minds able to think original thoughts, feel emotions, have concepts like love and beauty, and so on. It may seem that between them these two approaches pretty much cover the territory. Link them, and you have a complete answer to the question of mind.

However, neurophysiology and culture aren't independent: they are 'complicit'. By this we mean that they have evolved together, each changing the other repeatedly, and their mutual coevolution built on the unpredictable results of that ongoing interaction. The view of culture building on, and changing, brains is incomplete, because brains also build on, and change, culture. The concept of complicity captures this recursive, mutual influence.

We call the brain's internal capabilities 'intelligence'. It is convenient to give a similar name to all of the external influences, cultural or otherwise, that affect the evolution of the brain – and with it, the mind. We shall call these influences *extelligence*, a term that HEX has picked up thanks to once-and-future computing. Mind is not just intelligence *plus* extelligence – its inside and outside, so to speak. Instead, mind is a feedback loop in which intelligence influences extelligence, extelligence influences intelligence, and the combination transcends the capabilities of both.

Intelligence is the ability of the brain to process information. But intelligence is only part of what is needed to make a mind. And even intelligence is unlikely to evolve in isolation.

Culture is basically a collection of interacting minds. Without individual minds you can't have a culture. The converse is perhaps less obvious, but equally true: without a shared culture, the human mind cannot evolve. The reason is that there is nothing in the environment of the evolving mind that can drive it towards self-complication – becoming more sophisticated – unless that brain has something else fairly sophisticated to interact with. And the main sophisticated thing around to interact with is minds of other people. So the evolution of intelligence and that of extelligence are inextricably linked, and complicity between them is inevitable.

In the world around us are things that we, or other human beings, have created – things which play a similar role to intelligence but sit outside us. They are things like libraries, books, and the Internet – which from the viewpoint of extelligence would be better named the 'Extranet'. The Discworld concept of 'L-space' – library-space – is similar: *it's all one thing*. These influences, sources not just of information but of meaning, are 'cultural capital'. They are things that people put out into the culture, which can then sit there, or even reproduce, or interact in a way that individuals can't control.

The old artificial intelligence question: 'Can we create an intelligent machine?' viewed the machine as a once-off object in its own right. The problem, people assumed, was to get the machine's architecture right, and then program intelligent behaviour into it.

But that's probably the wrong approach. Of course, it is certainly *conceivable* that the collective extelligence of all the human beings interacting with that machine could put a mind into it – and in particular endow it with intelligence. But it seems much more likely that, unless you had a whole community of machines interacting with each other and evolving, providing the requisite extelligence too, then you wouldn't be actually able to structure the Ant Country of the neural connections of the machine in a way that could generate a mind. So the story of the mind is one of complicity and emergence. Indeed, mind is one of the great examples of complicity.

The internal story of the development of the mind can be summed up as a series of steps in which the key 'player' is the nerve cell. A nerve cell is an extended object that can send signals from one place to another. Once you've got nerve cells you can have networks of nerve cells; and once you've got networks, then a whole pile of stuff comes along free of charge. For example, there is an area of complexity theory called 'emergent computation'. It turns out that when you evolve a network – randomly chosen networks, arbitrary networks, not constructed with specific purposes – they do things. They do *something*, which may or may not seem meaningful; they do whatever it is that that network does. But you can often look at what that network does, and spot emergent features. You discover that even though its architecture was random, it evolved the ability to compute things. It carries out algorithmic processes (or something close to algorithmic processes). The ability to do calculations, computations, algorithms seems to come *free of charge* once you've invented devices that send signals from one place to another and react to those signals to send new signals. If you allow evolution you don't have to work hard to create the ability to do some kind of processing.

Once you've got that facility, it's a relatively short step to the ability to do specific kinds of processing that happen to be useful – that happen to offer survival value. All you need is the standard Darwinian selection procedure. Anything that's got that ability survives, anything that hasn't, doesn't. The ability to process incoming information in ways that extract an interesting feature of the outside world, react to it, and thereby make it easier to evade a predator or to spot food, gets reinforced. The brain's internal architecture comes from a phase space of possible structures, and evolution selects from that phase space. Put those two together and you can evolve structures in the brain that have specific functions. The brain's surroundings certainly influence the development of the brain.

Do animals have minds? They do to some extent, depending on the animal. Even simple animals can have surprisingly sophisticated mental abilities. One of the most surprising is a funny creature called a mantis shrimp.

It's like the shrimps you put inside a sandwich and eat, except that it's about 5 inches (12 cm) long and it's more complex. You can keep a mantis shrimp in a tank, as part of a miniature marine ecology. If you do, you'll find that mantis shrimps cause havoc. They tend to destroy things – but they also build things. One thing they love building is tunnels, which they then live in. The mantis shrimp is a bit of an architect, and it decorates the front of its tunnel with bits and pieces of things – especially bits and pieces of what it has just killed. Hunting trophies. It doesn't like to have just one tunnel – it's discovered that if you have one tunnel with one entrance, that's more correctly known as a 'trap'. So it likes to have a back entrance too – and more. By the time it's been in the tank for about two months, it's riddled the entire tank with tunnels, and you find it sticking its head out at one end or the other without seeing it pass between.

Years ago, Jack used to have a mantis shrimp called Dougal.* Jack and his students discovered that they could set Dougal puzzles. They would feed it shrimps and it would come out and grab the shrimp. Then they would put the shrimp inside a plastic container with a lid and after a little while Dougal would like to take the lid

* There was a television programme called *The Magic Roundabout*. One of the characters was a dog called Dougal, which looked a bit like a hairbrush. Mantis shrimps have the same general form, though not with hair.

off the container and eat the shrimp. And then they put an elastic band around the container to hold the lid on, and Dougal would learn to take the band off and open the container and eat the shrimp. And after a while if they stuck a shrimp in on its own, you could almost see the mantis shrimp coming out and looking disappointed: 'They haven't set me a puzzle, this is no fun, I don't want to play this game!' And it would take a long look at the shrimp and then go back into its tunnel without grabbing it.

Although we can think of no way to *prove* this, everyone got the strong impression that the shrimp was developing a little bit of a mind. Its brain had the potential to do so, and humans had provided it with the kind of context that would help it develop that potential. Wild mantis shrimps *don't* go out and play with elastic bands, because those aren't part of their environment, but if you give them that kind of stimulus, you change them. Because *we've* got minds, we also have the capacity to create a little bit of mind in a lot of other creatures.

Mind is a *process*, or a network of processes, going on inside the brain. It needs a certain amount of interaction with other minds in order to get anywhere. There isn't an evolutionary feedback loop that would train an incipient mind and make it develop *unless* it was getting somewhere. So where does such a loop occur? Human beings are part of a reproductive system – there are a lot of us, and we keep breeding new ones. In consequence, a large part of the environment of any human being is other human beings. In many ways this is the most important part of our environment, the part we respond to most deeply. We have all sorts of cultural systems, such as education, that exploit exactly this feature of our environment to develop the kind of mind that fits into the existing culture and helps to propagate it. So the context for an individual mind, as it evolves, is not that mind – it's lots of other minds. There is a complicit feedback loop between the entire collection of minds, and that of each individual.

Human beings have taken this process to such an extreme that part of that feedback loop has escaped from our control and is now outside us. In a sense, it has a mind of its own. This is extelligence, and we can't do without it. A lot of what makes us human is *not* passed on genetically – it is passed on culturally. It is passed on by the tribe, it is passed on through rituals, by teaching, by things that link brain to brain, mind to mind. Your genetics may make it possible for you to *do* this, it may make you better or worse at it than others, but genes don't actually encode the information that gets passed on. This process is the 'Make-a-Human-Being-Kit'. Each culture has devised a technique for putting into the minds of the next generation what it is that will make them put it into the minds of the generation after that – a recursive system that keeps the culture going. Lies-to-children often feature prominently.

We are running into problems doing this today, because old-style tribal cultures, even national cultures, are becoming intermingled with an international culture. This leads to clashes between what used to be separate cultures, triggering their breakdown. Go into any city in the world and you see adverts for Coca-Cola. Global commerce has put things into various cultures that are different from what they would have developed of their own accord. Coca-Cola does not have a huge influence on the Make-a-Human-Being-Kit, though, so it's acceptable to most cultures. On the whole, you don't find religious fundamentalists complaining about the existence of a Coca-Cola bottling factory in their country (well, you *do*, but generally because it's just a way of saying 'USA out!') However, if some fast-food chain in Islamic or Jewish countries was trying to sell porkburgers, there'd be plenty of protests.

Extelligence has become so powerful and so influential that nowadays one generation's culture may be radically different from the previous generation's culture. Second-generation immigrants often have an even worse problem, a culture clash. They've grown up in the 'new' country, and they've absorbed how that country works. They speak the language far more fluently than their parents ever can, but they've still got to *please* their parents. When they're at home, they have to behave in the manner of their original culture. But when they're at school, they have to live in the new culture. This makes them feel distinctly uncomfortable, and that can break the cultural feedback loop. Once the loop is broken, parts of the culture cease to be transmitted to the next generation: they drop out of the Make-a-Human-Being-Kit.

In this sense, extelligence is out of our control. It escaped our control when it became reproductive: extelligence being used to copy (bits of) extelligence.

The key step was the invention of printing. Prior to written language, extelligence was passed on by word of mouth. It still lived in people's minds: it was what the wise men and women of the village, the old people, knew. And all the while extelligence resided in human memories, it couldn't grow, because one person can remember only so much. When you could write things down, extelligence expanded a bit, but there is only so much that you can write down by hand. And it can't spread very far. So mostly you get things like the Egyptian monuments – the history of some particular ruler, his greatest battles, excerpts from the Book of the Dead ...

Another important but apparently mundane function of writing in human society is taxes, accounts, keeping track of property. These sound dull compared with the list of battles, but a growing society needs something better than an old man's memory of 'who owns what' and 'who paid how much'. The list was a great invention.

With printing came the possibility of disseminating information far more widely, and in quantity. Within a few years of printing becoming established in Europe there were fifty million books in existence, which means more books than people. Printing was a very slow procedure in those days, but nonetheless there were lots of printing presses, and you could sell whatever you printed, so there were plenty of pressures that encouraged printing to flourish. And then complicity really set in, because what's on a piece of paper can come back and bite you in the ankle. The rulers started putting constitutional rights and obligations down on paper, to protect their own position: once it's down on paper that the king has certain rights and obligations, then the paper can always be referred to later, and used as an argument.

But what the kings didn't realize, to start with, is that when they put their rights and obligations down on paper, they were implicitly constraining their own actions. *The citizens could read what was on the paper too.* They could tell if their king was suddenly assuming rights or obligations that were *not* on the piece of paper. The whole effect of law on human society started to change when you could write the law down, and anyone who could read could see what the law was. This didn't mean that the kings always *obeyed* the law, of course, but it meant that when they disobeyed it, everyone knew what they were doing. That had a big effect on the structure of human society. One minor aspect of it is that we always appear to be nervous of people who write things down …

At that point, extelligence and intelligence began to interact complicitly. Once an interaction becomes complicit, there's no way for an individual to control it. You can push things out into the extelligence, but you can't predict what influence they will have. What's out there is growing in a way that may be *mediated* by human beings, but – for example – the people printing books were largely printing them independently of their contents. Early on, *anything* in print would sell.

All words had power. But written words had a lot more. They still do.

So far we've talked as if extelligence is a single unified external thing. In some sense it is, but what is actually important is the interface between extelligence and the individual. This is a very personal feedback loop: we meet selections from extelligence through *our* parents, the books *we* read, the teachers who teach *us*, and so on. This is how the Make-a-Human-Being-Kit works, this is why we have cultural diversity. If we all responded to the same pool of extelligence in exactly the same way, we would all be the same. The whole system would suddenly become a kind of monoculture rather than a multiculture.

Human extelligence is currently going through a period of massive expansion. Much more is becoming *possible*. Your interface to extelligence used to be very predictable: your parents, teachers, relatives, friends, village, tribe. That allowed clusters of particular kinds of subculture to flourish, to some extent independently of the other subcultures, because you never got to *hear* about the others. Their world view was always filtered before it got to you. In *Whit*, Iain Banks describes a strange Scottish religious sect, and children who grow up in this sect. Even though some members of the sect are interacting with the outside world, the only *important* influences on them are what's going on within the sect. Even by the end of the story the character who has gone into the outside world and interacted with it in all sorts of ways has one idea in mind and one only – to become the leader of the sect and to continue propagating the sect's views. This behaviour is typical of human clusters – until extelligence intervenes.

Today's extelligence doesn't have a single world view, like a sect does. It doesn't really have a world view at all. Extelligence is becoming 'multiplex', a concept introduced by the science-fiction writer Samuel R. Delany in the novel *Empire Star*. Simplex minds have a single-world view and know exactly what everyone ought to do. Complex minds recognize the existence of different world views. Multiplex ones wonder how useful a specific world view actually is in a world of conflicting paradigms, but find a way to operate despite that.

Anyone who wants to can get on the Internet and construct a webpage about UFOs, telling everybody who accesses that page that UFOs exist – they're out there in space, they come down to Earth, they abduct people, they steal their babies ... They do all these things and it's absolutely definite, because *it's on the web*.

A prominent astronomer was giving a talk about life on other planets and the possibility of aliens. He made out the scientific case that somewhere out in the galaxy intelligent aliens *might* exist. At that point a member of the audience put his hand up and said 'we *know* they exist: it's all over the Internet.'

On the other hand, you can access another page on the Internet and get a completely different view. On the Internet, the full diversity of views is, or at least can be represented. It is quite democratic; the views of the stupid and credulous carry as much weight as the views of those who can read without moving their lips. If you think that the Holocaust didn't actually happen, and you can shout loud enough, and you can design a good web page, then you can be in there slugging it out with other people who believe that recorded history should have some kind of connection with reality.

We are having to cope with multiplexity. We're grappling with the problem right now: it's why global politics has suddenly become a lot more complicated than it used to be. Answers are in short supply, but one thing seems clear: rigid cultural fundamentalism isn't going to get us anywhere.

FORTY-FIVE

THE BLEAT
GOES ON

EXTELLIGENCE BLOOMED, faster than HEX could create extra space in which to apprehend it. It reached the seas and spread out across the continents, left the surface of the world, spun webs across the sky, reached the moon ... and went further, as intelligence sought things to be intelligent about.

Extelligence learned. Among many other things, it learned to fear.

The HEM filled up again as the wizards returned, unsteadily, from lunch.

'Ah, Rincewind,' said the Archchancellor. 'We're looking for a volunteer to go into the squash court and shut down the reactor, and we've found you. Well done.'

'Is it dangerous?' said Rincewind.

'That depends on how you define dangerous,' said Ridcully.

'Er ... liable to cause pain and an imminent cessation of respiration,' suggested Rincewind. 'A high risk of agony, a possible deficit of arms and legs, a terminal shortness of breath –'

Ridcully and Ponder went into a huddle. Rincewind heard them whispering. Then the Archchancellor turned, beaming.

'We've decided to come to a new definition,' he said. 'It is "not as dangerous as many other things". I beg you pardon ...' He leaned over as Ponder whispered urgently in his ear. 'Correction, "not as dangerous as *some* other things". There. I think that's clear.'

'Well, *yes*, you mean ... not as dangerous as some of the most dangerous things in the universe?'

'Yes, indeed. And among them, Rincewind, would be your refusal to go.' The Archchancellor walked over to the omniscope. 'Oh, another ice age,' he went on. 'Well, that *is* a surprise.'

Rincewind glanced at the Librarian, who shrugged. Only a few tens of thousands of years could have passed down there. The apes probably never knew what squashed them.

There was a lengthy rattle from HEX's write-out. Ponder walked over to read it.

'Er ... Archchancellor? HEX says he's found advanced intelligence on the planet.'

'Intelligent life? Down there? But the place is a snowball again!'

'Er ... not life, sir. Not exactly.'

'Hang on, what's this?' said the Dean.

There was, thin as a thread, a ring around the world. Spaced at regular distances were tiny dots, like beads, and from *them* more tiny lines descended towards the surface.

So did the wizards.

Wind howled across the tundra. The ice was only a few hundred miles away, even here at the equator.

The wizards faded into existence, and looked around them.

'What the *hell* happened here?' said Ridcully.

The landscape was a welter of scars and pits. Roads were visible where they had buckled up through the snow, and there were the ruins of what could only have been buildings. But half the horizon was filled with what looked very much like an etiolated version of one of the giant shellfish proposed by the Lecturer in Recent Runes. It must have been several miles across at the base, and extended upwards beyond the limit of vision.

'Did any of you do this?' said Ridcully accusingly.

'Oh, come *on*,' said the Dean. 'We don't even know what it is.'

Beyond the tangle of broken roadways the snow blew across deep trenches gouged out of the ground. Desolation reigned.

Ponder pointed towards the huge pyramid.

'Whatever we're looking for, it's in there,' he said.

The first thing the wizards noticed was the mournful bleating noise. It came and went in a regular way, on-off, on-off, and seemed to fill the entire structure.

The wizards wandered onwards, occasionally getting HEX to move them to different places. Nothing, they agreed, made much sense. The building was mostly full of roadways and loading docks, interspersed with massive pillars. It creaked, too, like an old galleon. They could hear the groaning noises, echoing far above. Occasionally, the ground trembled.

It was clear that important things happened in the centre. There were tubes, hundreds of feet high. The wizards recognized cranes, and failed to recognize huge engines of unknown purpose. Cables thick as a house rose into the darkness above.

Frost sparkled off everything.

Still the bleat went on.

'Look,' said Ponder.

Red words flashed on and off, high in the air.

'"A-L-A-A-M",' the Dean spelled out. 'I wonder why it's doing that? They seem to have invented magic, whoever they are. Getting letters to flash like that is quite difficult to do.'

Ponder disappeared for a moment, and then came back.

'HEX feels that this is a dumb-waiter,' he said. 'Er ... you know ... for lifting things to another level.'

'Going where?' said Ridcully.

'Er ... up, sir. Into that ... necklace around the world. HEX has been speaking to the intelligence here. It's a sort of HEX, sir. And it's nearly dead.'

'That's a shame,' said Ridcully. He sniffed. 'Where's everyone gone, then?'

'Er ... they made huge ... sort of ... big metal balls to live in. I know it sounds stupid, sir. But they've gone. Because of the ice. And there was a comet, too. Not very big. But it scared everyone. They built the ... the beanstalk things, and then they ... er mined metal out of floating rocks, and ... they left.'

'Where've they gone?'

'The ... intelligence isn't sure. It's forgotten. It says it's forgotten a lot.'

'Oh, I *understand*,' said the Dean, who'd been trying to follow this, 'Everyone's climbed up a great big beanstalk?'

'Er ... sort of, Dean,' said Ponder, in his diplomatic voice. 'In a manner of speaking.'

'Certainly messed the place up before they went,' said Ridcully.

Rincewind had been watching a rat scuttle away into the debris, but the words sunk in and exploded in his head.

'Messed up?' he growled. 'How?'

'Say again?' said Ridcully.

'Did you *see* the weather report for this world?' said Rincewind, waving his hands in the air. 'Two miles of ice, followed by a light shower of rocks, with outbreaks of choking fog for the next thousand years? There will be widespread vulcanism as half a continent's worth of magma lets go, followed by a period of mountain building? And that's *normal*.'

'Yes, well –'

'Oh, *yes*, there are some nice quiet periods, everything settles down, and then – whammo!'

'There's no need to get so excited –'

'I've *been* here!' said Rincewind. 'This is how this place *works*! And now, please, you tell me how, I mean *how*, can anything living on this world *possibly* mess it up? I mean, compared to what happens anyway?' He paused, and gulped air. 'I mean, don't get me wrong, if you pick the right time, yes, sure, it's a great world for a holiday, ten thousand years, even a few million if you're lucky with the weather but, good grief, it's just not a serious proposition for anything long term. It's a great place to grow up on, but you wouldn't want to *live* here. If anything's got off, the best of luck to them.'

He waved a finger at the rat, who was watching them suspiciously. Underneath them, the ground trembled again.

'See him?' he said. 'We *know* what's going to happen. In a million years or so his kids are going to be saying, wow, what a great world the Big Rat made for us. Or it'll be the turn of the jellyfish, or something that's still bobbing around under the sea that we don't even know about yet! There's no *future* here! No, that's wrong ... I mean there's always a future, but it belongs to someone else. You know what chalk's made of here? Dead animals! The actual *rock* is made of dead animals! There were some ...'

Even in his overheated state, he paused. It probably wasn't a good idea to remind people about the apes. A vague, suspicious guilt was nudging him.

'There were these creatures,' he said, 'and they were using limestone caves. Limestone's made from ancient blobs, I saw it being made, like snow in the water ... and these creatures are living in the bones of their ancestors! Really! This place ... this place is a kaleidoscope. You smash it up, wait a moment, and there's another pretty pattern. And another one. And another o ...' He stopped. And sagged. 'Could I have a glass of water, please?'

'That was a very ... interesting speech,' said Ponder.

'A point of view, certainly,' said Ridcully.

The other wizards had, however, lost interest. They usually did, if the speeches were not given by them.

'Shall I tell you something else?' said Rincewind, a little more calmly. 'This world is an anvil. *Everything* here is between a rock and a hard place. Every single thing on it is the descendant of creatures that have survived everything the world could throw at them. I just hope they never get angry ...'

The Senior Wrangler and the Dean had ambled towards a huge yellow cylinder. The word 'MAETNANS' was painted in large black letters on the side.

'Hey, you chaps!' the Dean shouted. 'There's something *talking* in here ...'

The inside of the cylinder reminded the wizards of a lighthouse. There was a spiral staircase; shaped cupboards lined the walls. Lights glowed dimly, whole constellations of them. Certainly the builders of this thing had discovered magic.

The 'A-L-A-A-M' word still blinked on and off in the air.

'I wish that wretched thing would stop,' said the Senior Wrangler.

The light vanished. The sound stopped.

'They've probably invented demons,' said the Dean airily. 'Listen ... hello.'

A pleasant female voice said, 'Elevator Unstable.'

'Oh, *magic*,' said Ridcully flatly. 'Well, we know how to deal with magic. We want to go up in the magic box, voice.'

'Do we?' said Ponder.

'Anything better than staying in this gloomy place,' said Ridcully. 'It'd be quite an interestin' experience, too. We'll take one last look the world and then, well ... frankly, that's it.'

'Instability Rising', said the voice. It did not sound worried by the news.

'What did it say?' said the Dean. 'Sounded like name of a place.'

'Very good, very good,' said Ridcully. 'Now let's be going shall we?'

The pattern of lights moved. Then the voice said, as if it'd been thinking it over, 'Emerjansi Override.'

The door slid shut. The cylinder jerked. Shortly afterwards, some pleasant music started, and didn't really get on anyone's nerves for several minutes.

The rat watched the thing rise up the cables in the centre of the pyramid.

The ground shook again.

Slowly, the web around the world came apart.

Ice walls had attacked some of the cable moorings on the ground, but instability was already there, working inexorably as it had done for the past few weeks, turning little movements into big movements.

Slowly, one cable broke free from its pyramid, glowing red-hot as it was jerked through the atmosphere, flailing across the sky.

Around the curve of the world, the others danced and groaned ...

When the end finally came, it took only a day. The lines folded around the centre of the world, writhing incandescently across hundreds of miles of snow. The necklace tore apart far above. Some bits drifted away. Others spun gently towards the surface, to impact hours later.

A ring of fire burned for a while around the equator.

And then the cold returned.

As the wizards said, it would all be the same in a hundred million years' time. But it would be different tomorrow.

In the deserted High Energy Building, HEX turned the omniscope outwards, homing in on signs of the strange new life.

It found comet cores, strung on cables thousands of miles long. There were dozens of these trains, many millions of miles from the frozen world, accelerating into the abyss between the stars.

Lights twinkled on their surfaces. The extelligence inside appeared to be travelling hopefully.

A yellow cylinder tumbled gently across the darkness.

It was empty.

FORTY-SIX
WAYS TO LEAVE YOUR PLANET

RINCEWIND'S IMPASSIONED SPEECH HAS A POINT. If you think he's overstating his case, and that the Earth is really an idyllic place to live, bear in mind that he's been on our planet a lot longer than we have, and he's seen a lot that we've missed, because we experience the world on a much shorter timescale than the wizards have done. We think the planet's a great place. We grew up here. We were made for it, and it's just right for *us* ... at the moment.

Tell that to the dinosaurs.

You can't, can you. That's the point.

We're not suggesting that you sell up everything and start building a lifeboat. But even the United States congress is beginning to wonder just how safe our planet really is, and politicians are not usually known for taking long-term views. The sight of Shoemaker-Levy 9 smashing into Jupiter raised a few political eyebrows. Tentative schemes are afoot to set up a defence system against incoming comets and asteroids. Spotting them early enough is the trick. Find them quickly, and a modest little rocket motor can save our planetary bacon.

It is in many ways amazing that life on Earth has survived everything that the universe has so far thrown at it. Evolution runs on Deep Time – less than a hundred million years hardly counts. Life is extremely resilient, but individual species are not. They last a few million years and then they become obsolete. Life persists by changing – by being a series of opening chapters. But, being human, we'd like to see our own story turn into at least a blockbuster dekalogy.

We can take small comfort in one thing. Although right now we don't worry enough about incoming disaster from Up There, we do worry a lot about home-grown disaster Down Here: nuclear warfare, biological warfare, global warming, pollution, overpopulation, destruction of habitat, burning of the rainforests, and so on. However, there's no danger that human actions will wipe out *the planet*. Compared to what nature has already done, *and will do again*, our activities barely show up. One large meteorite packs more explosive power than all human wars put together, a hypothetical World War III included. One Ice Age changes the climate more than a civilization's worth of carbon dioxide from car exhausts. As for something like the Deccan Traps ... you wouldn't want to *know* how nasty the atmosphere could become.

No, we can't destroy the Earth. We *can* destroy ourselves.

No one would care. The cockroaches and the rats will come back, or if the worst comes to the worst the bacteria miles below ground will start to write a new opening chapter in the Book of Life. Someone else will read it.

If we really deserve the name *Homo sapiens*, then we can do at least two things to improve our chances. First, we can learn to manage our impact on the environment. The fact that nature deals the occasional death blow doesn't hand us an excuse to imitate it. *We* invented ethics. Our environment is sufficiently buffeted by various forces that the last thing it needs is humanity throwing extra spanners in the works. At the most selfish level, we might be buying ourselves some time.

We could use that time to put some of our eggs in another basket.

One of the great dreams of humanity has been to visit other worlds. It's starting to look as though this might be a very good idea – not just for fun and profit, but for survival.

We'd better say right now that none of this is science fiction. Or, rather, yes, it is science fiction, it's the very stuff of science fiction, because some of the best science-fiction writers (you don't see their stuff on TV) have been dealing with it for many decades. But that does not mean it's not *real*. Ices Ages happen. Big, big rocks come screaming out of the sky, and you need rather more than Bruce Willis flying the Space Shuttle as if it was the Millennium Falcon to stop them.

Our urge to explore the universe may be just another case of monkey curiosity, but there seems to be a deep impulse that urges us to find new lands to map and new worlds to conquer. Maybe there's an inbuilt urge to spread out – one leopard can't eat *all* of you if you spread out.

It is an urge that has driven us into every corner and crevice of our own planet, from the ice-floes of the Arctic to the deserts of Namibia, from the depths of the Mariana Trench to the peak of Everest. Most of us incline to Rincewind's view of a comfortable lifestyle and much prefer to stay at home, but a few are too restless to be happy anywhere for very long. The combination is a powerful one, and it has shaped our species into something very unusual, with collective capabilities beyond the understanding of any individual. We may not always use that combination *wisely*, but without it we would be greatly diminished. And it's offering a real opportunity.

Even a dream can work miracles. When Columbus (re-)discovered America, and Europe found out that it existed, he was looking for a new route to the Indies. He had convinced himself – on grounds that most scholars at the time found totally spurious – that the Earth was considerably smaller than was generally thought. He calculated that a relatively short voyage westward, from Africa, would lead to Japan and India. The scholars were right, Columbus was wrong – but it is Columbus that we remember, because he made the world smaller. He had the courage to set sail into an empty sea, sustained only by the belief that there was something important on the other side.

At least we can *see* where we ought to go. Columbus had to back a hunch.

A dirty great Saturn-V rocket with a tiny Apollo capsule on top was the first practical method for getting out of the Earth's gravity well altogether. By this we don't mean that the Earth's gravitational pull becomes zero if you go far enough away, which is a common misconception: we mean that if you go fast enough, then the Earth's gravity can never pull you back down. Celestial mechanics operates in the phase space of distance *and* velocity, its 'landscape' involves speeds as well as lengths. Only when we understood enough about gravity and dynamics to appreciate this point did we stand any chance of making technology like Apollo work.

You can see this clearly from earlier suggestions, which were imaginative – in an earthbound sort of way – but fantastic and impractical, at least on Roundworld. In 1648 Bishop John Wilkins listed four possible ways to leave the ground: enlist the aid of spirits or angels, get a lift from birds, fasten wings to your body, or build a flying chariot. If we wanted to be charitable, we could interpret the last two as aircraft and rockets, but Wilkins was clearly unaware that the Earth's atmosphere doesn't extend all the way to the Moon. A sixteenth-century engraving by Hans Schaüffelein depicts Alexander the Great carried into space by two griffins – no noticeable improvement. Bernard Zamagna conceived of an aerial boat, and others suggested the use of balloons.

Every age fantasized about technology that already existed. In Jules Verne's *From the Earth to the Moon* of 1865 the journey was accomplished by firing a space capsule from a huge gun in Florida; its 1870 sequel *Around the Moon* involved a series of such capsules, forming a space train. Verne got Florida right – he knew that the Earth's spin produces centrifugal force, which helps the capsule to leave the planet more easily, and he knew that this force was greatest at the equator. Since the protagonists in his book were American, Florida was the best bet. When NASA started launching rockets, it came to the same conclusion, and the space facility at Cape Canaveral was born.

Big guns have deficiencies, such as a tendency to laminate passengers to the floor because of rapid acceleration, but modern technology does make it possible to avoid this by applying the acceleration gradually. Rockets are currently more practical from the engineering point of view, though that could change. In 1926 Robert Goddard invented the liquid fuel rocket. The first one rose to the dizzy height of 40 feet (12.5 m). Rockets have come a long way since then, taking men to the Moon and instruments to the edge of the solar system. And they are much better rockets. Even so, there's something ... *inelegant* about heading off the planet on a giant disposable firework.

Until recently, there has been a general assumption that the energy to get into space has to be carried with the craft. However, we already have the beginnings of one way to get off the Earth that keeps the power source firmly on the ground. This is laser propulsion, in which a powerful beam of coherent light is aimed at a solid object and literally pushes it along. It takes a lot of power, but prototypes invented by Leik Myrabo have already been tested at the High Energy Laser System Test Facility at White Sands. In November 1997 a small projectile reached a height of 50 feet (15 m) in 5.5 seconds; by December this had been improved to 60 feet (20 m) in 4.9 seconds. This may not sound impressive, but compare with Goddard's first rocket. The method involves spinning the projectile at 6000 revolutions per minute to achieve gyroscopic stability. Then 20 laser pulses per second are directed towards a specially shaped cavity, heating the air beneath the craft and creating a pressure wave of thousands of atmospheres with temperatures up to 30,000° Kelvin – and that's what propels the projectile. At higher altitudes the air becomes very thin, and a similar craft would need an onboard fuel source. Fuel would be pumped into the cavity to be vapourized by the laser. A megawatt laser could lift a 2-pound (1 kg) craft into orbit.

It is also a very powerful weapon ...

Another possibility is power beaming. It is possible to 'beam' electromagnetic power from the ground in the form of microwaves. This isn't just fantasy: in 1975 Dick Dickinson and William Brown beamed 30 kilowatts of power – enough for thirty electric fires – over a distance of one mile. James Benford and Myrabo have suggested launching a spacecraft using millimetre range microwaves which are not attenuated by the atmosphere. This is a variation on the laser method and would use the same kind of projectile.

Both of these methods rely on a lot of raw power, betraying traces of the basic engineering assumption that getting into space needs a lot of energy to overcome the Earth's gravity. They do have the advantage that the raw power is just sitting on the planet; the

1,000 megawatt power station your laser launcher would require could generate for the National Grid when a launch wasn't going on.

A method of greater subtlety is the bolas, first proposed in the 1950s. Traditionally, a bolas is a hunting device made by tying three weights to strings and then tying the ends of the strings together. When thrown, it spins, pulling the weights apart, until the strings hit the target, at which point the weights spiral rapidly inwards and deal a killing blow. The same sort of device could be set up in a vertical plane above the equator, a bit like a giant ferris wheel with only three spokes. On the ends of the spokes would be pressurized cabins. The lowest part of the bolas's swing would be somewhere in the lower atmosphere, the top part way out in space. You would fly up in an aircraft, transfer to the first passing cabin, and be whisked skywards. The biggest obstacle to making such a machine is the cable, which has to be stronger than any known material – but carbon fibre is well on the way to combining enough strength with enough lightness. Friction with the atmosphere would gradually slow the bolas's rotation down, but that could be compensated for using solar power arrays up in space.

The most celebrated device of this type, however, is the space elevator. We discussed this earlier, both as a serious technological idea and as a metaphor: here we give a few more details. In essence, the space elevator starts out as a satellite in geosynchronous orbit. Then you drop a cable from it to the ground, and the rest is a matter of building a suitable cabin and, again, finding suitable material for the cable. You get the material up there using rockets or a whole cascade of bolases (and once you've got a small cable you can haul up the stuff for the bigger one). You only need to do all this *once*, so the cost is irrelevant over the longer term.

As we emphasized at the start of the book, once there is as much traffic is coming down as is going up, getting off the ground is essentially free and requires zero energy. At that point you build your interplanetary spacecraft up in space, using raw materials from the Moon or the asteroid belt. So the space elevator gives you *a new place to start from* – which is why we've used it as a metaphor for processes like life.

The idea of a space elevator was originated by the Leningrad engineer Y.N. Artsutanov in 1960, in an article in *Pravda*. He called it a 'heavenly funicular' and calculated that it could lift 12,000 tons per day into orbit. The idea came to the attention of Western scientists in 1966, thanks to John Isaacs, Hugh Bradner, and George Backus. These scientists weren't interested in getting into space: they were oceanographers – the only people seriously interested in hanging things on long cables. Except that they wanted to hang them down into the ocean bottoms, not up into space. The oceanographers were unaware of the earlier Russian work, but Artsutanov's anticipation quickly became known to Western scientists too. The astronaut and artist Alexei Leonov published a painting of a space elevator in action in 1967.

Such a simple but mostly impractical idea is likely to occur to lots of people, but wouldn't become widely known *because* it looks impractical with current technology, and that means that it will be re-invented independently by many people. In 1963 the science-fiction author Arthur C. Clarke considered suspending a lower satellite by cable from a geosynchronous one, as a way to increase the number of effectively geo- synchronous satellites for communication purposes. Later he realized that the same method would lead to the space elevator, an idea that he developed in his novel *The Fountains of Paradise*. In 1969 A.R. Collar and J.W. Flower also considered suspending a lower satellite by cable from a geosynchronous one And in 1975 Jerome Pearson suggested an 'orbital tower' that was essentially the same idea.

You can, of course, suspend more than one cable – once you've got *one* space elevator you can lift everything else that you need into space at low cost, so why not go the whole hog? Charles Sheffield's *The Web Between the Worlds* envisages a whole ring of space elevators round the equator. This is what the wizards have found. Ironically, because human civilization has taken such a short time to develop, on evolutionary timescales, the wizards missed us …

Is the space elevator anything other than a wild fantasy? Could one really be built in the near future? In 2001 two NASA teams carried out feasibility studies, and both concluded that the space elevator is technologically possible. Just. David Smitherman, who led one team, reckons it could be in place by 2100.

The main problem is the cable. The tension in the cable is lowest near the ground, and highest at the top, because each section of cable has to support only the weight of cable below it. So the cable should be made thin at the bottom, and thicker towards the top. The big question is: which material has enough tensile strength? Steel won't do: a steel cable 4 inches (10 cm) wide at the bottom would have to be 2.5 trillion miles (4 trillion km) across at the top. (This is an engineer's way of saying 'don't use steel': it's too heavy and the stress goes up extremely fast as the cable gets longer.) Kevlar would be more practical: the top would then need to be only 1600 metres across – just over a mile wide. But even this is not practical enough.

For the size to be acceptable, the cables tensile strength needs to be at least 62.5 gigapascals – 30 times stronger than steel and 17 times stronger than Kevlar. Such materials do exist: the best known is the carbon nanotube, a molecule of carbon shaped like a hollow cylinder and related to the famous molecule buckminsterfullerene, which is made from 60 atoms of carbon and is shaped like a soccer ball. The tensile strength of a carbon nanotube is at least 130 gigapascals, more than twice as strong as necessary. The only snag is, that right now the longest carbon nanotubes we can make extend for only a few millionths of a metre. But if that could be increased to 4 millimetres, then the nanotubes could be embedded in a composite material with the necessary strength.

A second problem is the base. The higher the bottom of the cable is off the ground, the more material is saved at the top, where most of the mass is. This is why the cable in our story has a huge 'etiolated whelk' at its base. The NASA study concluded that a tower at least 6 miles (10 km) high would be best. It could be built on a mountaintop, to reduce the height needed, but if the cable were to snap, the main debris would then fall on land. So a tower in the ocean at the equator would make more sense. Current construction methods could, in principle, build a 12-mile (20-km) tower.

The final design problem is how to transport capsules up and down the cable. Whatever method is used, it has to be low on maintenance and high on speed. Magnetic levitation looks good.

After that, it's mostly a question of protecting the cable against meteor-strikes and incoming high-energy particles. A piece of cake.

Having built your space elevator, you're now in a position to colonize other worlds. The obvious first destination is Mars. You get there in a cloud of small, mass-produced ships, and once you've got there one of the first things you do is drop down a cable and build a Martian space elevator. You're up in orbit anyway, so why not take advantage of the fact? Again, this is the metaphorical aspect of the space elevator: as soon as just one exists, it opens up a vast range of new possibilities. However, you'll probably need to land a team by some other method in order to construct the complex at the bottom to which the cable will be tethered.

Mars isn't a great place to live, so the next step is to terraform it – to make it more earthlike. There are reasonably plausible methods for doing that, detailed at length in Kim Stanley Robinson's series *Red Mars, Green Mars, Blue Mars*. Mars is no improvement when it comes to meteor-strikes, but at least the colony on Mars is unlikely to get wiped out at the same time as the main population on Earth. Because life is reproductive, if one of them *does* get wiped out, it can quickly be re-colonized from the other. After a few centuries, you'd hardly notice any difference. Still, it may be better to be more ambitious and go to the stars. By the time we're ready for that, we'll have interferometer telescopes good enough to spot which stars have suitable planets. The only problem, then, will be to get there.

There are plenty of suggestions, and we won't add to them. Think of mid-Victorians predicting life in the 1990s. The dynamic of extelligence is emergent or, to put it another way, we haven't the faintest idea what we'll think of next but it'll probably surprise us.

One way, if all else fails, is the Generation Ship – a huge vessel that can hold an entire city of people, who live, breed, educate, and die throughout the centuries-long journey. Make it big and interesting enough, and they may even lose interest in the destination.

The Discworld almost counts as one of these; it's on a journey, the inhabitants don't know where they're going, the designers have given it a small controllable sun (thus doing away with all those nasty fluctuations) and no less than five bio-engineered creatures positively *delight* in clearing local space of intrusive debris ...

Back on our world, you could take a *really* long-term view and seed the galaxy with genetically engineered bacteria, carefully tailored so that whenever they find a suitable planet they eventually evolve into humanoid life (or life, at least). We would die out, but maybe our fleet of cheap, slow ships might seed a few new Earths somewhere.

There's no shortage of ideas. Some might even be practical. The galaxy beckons. We might die trying – but since we're going to die anyway, why not try?

And what will we find out there? Will we find a radically different kind of 'space elevator', for instance? Well, if there are aliens that live on neutron stars, as Robert L. Forward describes in *Dragon's Egg*, then they might escape by tilting their world's magnetic axis, turning it into a pulsar, and surfing its plasma jet. Perhaps all those pulsars were formed in this way. Like any 'space elevator', if you can manage the trick once, the rest is easy. The inhabitants of one neutron star managed it, and colonized all the others, founding the Pulsar Empire ...

And since we can envisage new kinds of physical space elevator, there must surely also be new kinds of metaphorical space elevator. Not just aliens a bit like us, but radically different new kinds of life.

What else could live on a neutron star?

They're waiting.

FORTY-SEVEN

YOU NEED
CHELONIUM

'THAT,' SAID THE DEAN, 'was a very unpleasant business. Good thing we weren't really there.'

Rincewind was sitting at the end of the long table, his chin on his hand.

'Really?' he said. 'You thought that was bad? Try having a comet land on you. That really makes your day.'

'It was the music that really got on *my* nerves,' said the Senior Wrangler.

'Oh, *well*, good job the planet's a snowball, then,' said Rincewind.

'I call this meeting to order,' said Ridcully, thumping the table. 'Where's the Bursar?'

The wizards looked around the main hall of the High Energy Magic building.

'I saw him half an hour ago,' the Dean volunteered.

'We are quorate, nevertheless,' said Ridcully. 'Now ... the magic flux is almost run down, although HEX reports that the model universe appears to be continuing on internal power. Amazing the way the whole place seems to strive to keep existing. However ... gentlemen, the project is at an end. All it is has taught us is that you can't make a world out of bits and pieces. You need chelonium for a *proper* world. And you certainly need narrativium, otherwise the life you get is a lot of opening chapters. A comet is no way to end a story. Ice and fire ... that's *very* primitive.'

'Poor old crabs,' said the Senior Wrangler.

'Goodbye, lizards,' said the Dean.

'Farewell, my limpet,' said the Lecturer in Recent Runes.

'What were the ones that left?' said Ponder.

'Er ...' said Rincewind.

'Yes?' said the Archchancellor.

'Oh, nothing. I had a thought … but it couldn't possibly work.'

'Some of the bears seemed quite bright,' said Ridcully, who had naturally sided with a lifeform that resembled him in several particulars.

'Yes, yes, it was probably the bears,' said Rincewind quickly.

'We couldn't watch the whole world all the time,' said Ponder. 'Something could have evolved quickly, I suppose.'

'Yes, that's right, something probably evolved quickly,' said Rincewind. 'I shouldn't think there was any unauthorized interference in any way.'

'Good luck to them, whatever shape they're in,' said Ridcully. He assembled his papers. 'That's it, then. I won't say it hasn't been an interesting few days, but reality calls. Yes, Rincewind?'

'What are we going to do with the snow globe – I mean, the world?' said Rincewind.

As one wizard, they looked across at the world spinning gently in its dome.

'Is it any use to us, Mister Stibbons?' said Ridcully.

'As a curiosity, sir.'

'This university is *stuffed* with curiosities, young man.'

'Well, then … only as very large paperweight.'

'Ah. Rincewind … you *are* the Professor of Cruel and Usual Geography, so I suppose this is *right* up your street –'

There was a rattle from HEX's tray. Ponder pulled out the paper. It said: +++ The Project Must Be Kept Safe +++

'Fine. Rincewind can put it on a high shelf so that it doesn't get knocked,' said Ridcully, rubbing his hands together.

+++ Recursion Is Occurring +++

Ridcully blinked at the writing.

'Is that a problem?'

HEX creaked. There was a flurry of activity in the ant tubes, Eventually the write-out clattered for some time.

Ponder picked up the message.

'Er … it's addressed to Mrs Whitlow,' he said. Er … it's rather odd …'

Ridcully looked over his shoulder.

'"Don't Dust It",' he read.

'She's a devil with a duster,' said the Senior Wrangler. 'The Dean nails his door shut when he leaves his study.'

The write-out clattered again.

'"This Is Important",' Ponder read.

'Not a problem, not a problem,' said Ridcully. 'So on to the next item. Ah, yes. We have to shut down the reacting engine. No, don't get up, Rincewind, I've had the door locked. The interior of the squash court is still just a tiny bit not entirely completely safe, is that right, Mr Stibbons?'

'Very definitely!'

'And therefore the area within it quite clearly counts as –'

'Let me guess,' said Rincewind. 'It's cruel and unusual geography, yes?'

'Well, done, that man! And all you have to do –'

A sound that *had* been on the limit of hearing suddenly descended through the scales. And there was silence.

'What's that?' said Ridcully.

'Nothing,' said Rincewind, with unusual accuracy.

'The reacting engine has shut down,' said Ponder.

'By itself?'

'Not unless it can pull its own levers, no …'

The wizards clustered around the door to the old squash court. Ponder held up his thaumometer.

'There's hardly any flux now,' he said. 'It's practically background … Stand back …'

He opened the door.

A couple of white pigeons flew out, followed by a billiard ball. Ponder pulled aside a cluster of flags of all nations.

'Just natural fallout,' he called out. 'Oh …'

The Bursar ambled around the side of the reacting engine, waving a squash racket.

'Ah, Ponder,' he said. 'Have you wondered if Time isn't simply Space rotated through a right angle?'

'Er … no …' said Ponder, watching the man carefully for signs of thaumic breakdown.

'It would certainly make pretzels very interesting, don't you think?'

'Er ... have you been playing squash, sir?' said Ponder.

'You know, I'm really coming to believe that a closed contour is a boundary, up to parametrization, if and only if it is homotopic to zero,' said the Bursar. 'And, for preference, coloured green.'

'Did you touch any switches, sir?' said Ponder, maintaining a careful distance.

'This thingy here does make some shots very difficult,' said the Bursar, hitting the reacting engine. 'I was trying to hit the rear wall around last Wednesday.'

'I think perhaps we should leave,' said Ponder in a clear, firm tone. 'It will soon be teatime. There will be jelly,' he added.

'Ah, the fifth form of matter,' said the Bursar brightly, following Ponder.

The other wizards were waiting just outside the door.

'Is he all right?' said Ridcully. 'I mean by general bursarial standards, of course.'

'It's hard to tell,' said Ponder, as the Bursar beamed at them. 'I think so. But the reacting engine must had been putting out quite a high flux when he went in.'

'Perhaps none of the thaumic particles hit him?' said the Senior Wrangler.

'But there's millions of them, sir, and they can pass through anything!'

Ridcully slapped the Bursar on the back.

'Bit of luck for you, eh, Bursar?'

The Bursar looked puzzled for a moment, and then vanished.

FORTY-EIGHT

EDEN AND CAMELOT

 THIS BOOK WASN'T CALLED The *Religion of Discworld* for a reason, although – Heaven knows – there is plenty of raw material.* All religions are true, for a given value of 'truth'.

The disciplines of science, however, tell us that we live on a world formed from interstellar debris some four billion years ago in a universe which itself is about 15 billion years old (which is science-speak for 'a very long time'); that in the ensuing years it has been pummelled and frozen and re-arranged on a regular basis; that despite or rather *because* of this, life turned up very quickly and seems to spring back renewed and re-formed from every blow; and that we ourselves evolved on this planet and, with the suddenness of a bursting dam, became Top Species in a very short period of time.

Actually, science tells us that many cockroaches, bacteria, beetles, and even small mammals might argue that last statement, but since they are not good at debate and can't speak, who cares what they think? Especially since they can't, eh? A key thing about big brains is this: they know big brains are good.

Most of us don't think like scientists. We think like the wizards of Discworld. Everything in the past was leading inevitably to Now, which is the important time.

While the news that the Earth is a small planet in a dull part of the universe has caught on in recent centuries, it's only in the last few decades that the words 'the Earth' have come to mean, for a significant proportion of any society, 'the planet' rather than 'the soil'.

It was probably those photographs of the Earth seen from the Moon that did it. We saw the whole planet as a single *thing*, rather

* Explained to the hilt in *The Science of Discworld II: The Globe.*

than just the bit of it that we were standing on. And it looked fragile, and kind of lonely…

We watch the fireworks as great balls of ice plummet into the
atmosphere of a nearby planet and, although any one of them would
have *seriously* troubled the Earth, the event was just that: a firework
display. As one old lady told a news reporter, 'that sort of thing happens in Outer Space'. But we're in Outer Space, too, and it might
pay us to get good at it.

The dinosaurs were not, as suggested in *Jurassic Park*, 'selected
for extinction' – they were clobbered by a very large rock, and/or
its after-effects. Rocks don't think.

The dinosaurs were in fact doing very well, and had merely neglected to develop three-mile thick armour plating. They *may* even
have evolved something that we'd recognize as 'early civilization';
we shouldn't underestimate how much the surface of the planet can
change in 65 million years. But rocks don't care, either.

But even if the rock had missed, there were other rocks. And if
they had missed too, then we should be aware that the planet has
other, home-grown means of disposal.

Evidence is emerging that suggests that other extinctions were
caused by 'natural' but catastrophic changes in the planet's atmosphere. A case is being made that indicates that the very *existence* of
life on Earth will, periodically, trip a catastrophe.

Rocks don't *mind*.

This will probably not happen tomorrow. But, one day, it will. And
then Rincewind's kaleidoscope is shaken up for a *new* pretty pattern.

Eden and Camelot, the wondrous garden-worlds of myth and
legend, are here *now*. This is about as good as it ever gets. Mostly,
it's a lot worse. And it won't stay like this for very long.

There are, perhaps, choices. We could leave. We've dealt with
that. Considerably optimism is required. But there *might* be other
small blue planets out there … By definition, though, Earthlike
worlds will have life on them. That's *why* they'll be Earthlike. And
the trouble is that the more Earthlike it is, the more troublesome it
would be. Don't worry about the laser-wielding monsters – you can
talk to them, if only about lasers. The real problem is more likely to

be something very, very small. In the morning you get a rash. In the afternoon, your legs explode.*

The other 'choice' is to stay. We *may* be lucky – we tend to be. But we won't be lucky forever. The average life of a species is about five million years. Depending on how you define humanity, we may already be close to the average.

A useful project, and one that's much cheaper to achieve, is to leave a note to the next occupiers, even if it is only to say 'We Were Here'. It may be of interest to a future species that even if they are alone in space, they're not alone in Time.

We may already have left our marker. It depends on how long things will *really* last on the Moon, and if, in a hundred million years, anyone else feels it necessary to go there. If they do, they may find the abandoned descent stages of the Apollo Moon landers. And they'll wonder what a 'Richard M. Nixon' was.

How much luckier are the inhabitants of Discworld. They *know* they live on a world made for people. With a large hungry turtle, not to mention the four elephants, interstellar debris becomes lunch rather than catastrophe. Large-scale extinction has more to do with magical interference than random rocks or built-in fluctuations; it may have the same effect, but at least there is someone to *blame*.

Unfortunately, it does reduce the scope for asking interesting questions. Most of them have already been answered. Certainty rules. Mustrum Ridcully is not the kind of person who would tolerate an Uncertainty Principle, after all.

Back in Roundworld, there is perhaps one point worth making.

Just suppose there is *nothing else*. Arguments about intelligent life on other worlds have always been highly biased by the desires of those doing the arguing that there should *be* intelligent life on other worlds, and we three are among them. But the argument is a house of cards with no card on the bottom. We know of life on one world.

---

* This is probably another lie. Alien microbes are unlikely to find us edible. So are alien tigers, although they might do us quite a lot of damage in finding out. But certainly an alien world will have a whole host of nasty surprises, if we are not very careful. We can't tell you what they'll be. They'll be a *surprise*.

Everything else is guesswork and naked statistics. Life may be so common through the universe that even the atmosphere of Jupiter is alive with Jovian gasbags and every cometary nucleus is home to colonies of microscopic blobules. Or there may be nothing alive at all, anywhere else but here.

Perhaps intelligent life arose before humanity, and perhaps it will again when humanity's span has become a rather complex layer in the strata. We can't tell. Time does not simply, as the hymn says, bear all its sons away – it can easily see the disappearance of the entire continent on which they stood.

In short, in a universe a billion Grandfathers long and a trillion Grandfathers wide, there may be just a few hundred thousand years on one planet where a species worried about something other than sex, survival, and the next meal.

This is *our* Discworld. In its little cup of spacetime, humanity has invented gods,* philosophies, ethical systems, politics, an unfeasible number of ice-cream flavours and even more esoteric things like 'natural justice' and 'boredom'. Should it matter to us if tigers are made extinct and the last orangutan dies in a zoo? After all, blind forces have repeatedly erased species that were probably more beautiful and worthy.

But we feel it *does* matter, because humans invented the concept of things 'mattering'. We feel we ought to be brighter than a mile of incandescent rock and a continent-sized glacier. Humans seem to have created, independently, in many places and at various times, a Make-a-Real-Human-Being Kit, which begins with prohibitions about killing and theft and incest and is now groping towards our responsibilities to a natural world in which, despite its ability to hurt us mightily, we nevertheless have a godlike power.†

We advance arguments about saving rainforests because 'there may be undiscovered cancer cures in there', but this is because extelligence wants to save rainforests and the cancer-cure argument might convince the bean-counters and the fearful. It might have a

---

* We apologize to any *real* gods.

† Unfortunately, huge malicious destructive force is a god-like power.

real basis in fact, too, but the real reason is that we feel that a world with tigers and orangutans and rainforests and even small unobtrusive snails in it is a more healthy and interesting world for humans (and, of course, the tigers and orangutans and snails) and that a world without them would be dangerous territory. In other words, trusting the instincts that up until now have generally seen us through, we think that Tigers Are Nice (or, at least, Tigers Are Nice In Moderation And At A Safe Distance).

It's a circular argument, but in our little round human world we've managed to live on circular arguments for millennia. And who else is going to argue with us?

FORTY-NINE

AS ABOVE,
SO BELOW

'RINCEWIND WALKED VERY GINGERLY towards his office, the globe of the project held carefully in his hands.

He would have expected an entire universe to be heavier, but this one seemed on the light side. It was probably all that space.

The Archchancellor had explained at length to him that although he would be *called* the Egregious Professor of Cruel and Unusual Geography, this was only because that was cheaper than repainting the title on the door. He was not entitled to wages, or to teach, or express any opinions on anything, or order anyone around, or wear any special robes, or publish anything. But he could turn up for meals, provided he ate quietly.

To Rincewind, it sounded like heaven.

The Bursar appeared right in front of him. One moment there was an empty corridor, the next moment there was a bemused wizard.

They collided. The sphere went up in the air, turning gently.

Rincewind rebounded from the Bursar, looked up at the ball curving through the air, flung himself forward and down with rib-scraping force and caught it a few inches from the stone floor.

'Rincewind! Don't tell him who he is!'

Rincewind rolled over, clasping the little universe, and looked back along the passage. Ridcully and the other wizards were advancing slowly and cautiously. Ponder Stibbons was waving a spoonful of jelly invitingly.

Rincewind glanced up the Bursar, who was looking perplexed.

'But he's the Bursar, isn't he?' he said.

The Bursar smiled, looked puzzled for a moment, and vanished with a 'pop'.

'Seven seconds!' shouted Ponder, dropping the spoon and pulling out a notebook. 'That'll put him in … yes, the laundry room!'

The wizards hurried off, except for the Senior Wrangler, who was rolling a cigarette.

'What happened to the Bursar?' said Rincewind, getting to his feet.

'Oh, young Stibbons reckons he's caught Uncertainty,' said the Senior Wrangler, licking the paper. 'As soon as his body remembers what it's called it forgets where it's supposed to be.' He stuck the bent and wretched cylinder in his mouth and fumbled for his matches. 'Just another day at Unseen University, really.'

He wandered off, coughing.

Rincewind carried the sphere though the maze of dank passages and into his office, where he cleared a space for it on a shelf.

The ice age had cleared up. He wondered what was happening down there, what gastropod or mammal or lizard was even now winding up its elastic ready to propel itself towards the crown of the world. Soon, without a doubt, some creature would suddenly develop an unnecessarily large brain and be forced to do things with it. And it'd look around and probably declare how marvellous it was that the universe had been built to bring forward the inevitable development of creature-kind.

Boy, was it in for a shock …

'Okay, you can come out,' he said. 'They've lost interest.'

The Librarian was hiding behind a chair. The orangutan took university discipline seriously, even though he was capable of clapping someone on both ears and forcing his brain down his nose.

'They're busy trying to catch the Bursar right now,' said Rincewind. 'Anyway, I'm sure it couldn't have been the apes. No offence, but they didn't look the right sort to me.'

'Ook!'

'It was probably something out of the sea somewhere. I'm sure we didn't see most of what was going on.'

Rincewind huffed on the surface of the globe, and polished it with his sleeve. 'What's recursion?' he said.

The Librarian gave a very expansive shrug.

'It looks okay to me,' said Rincewind. 'I wondered if it was some sort of disease ...'

He slapped the Librarian on the back, raising a cloud of dust. 'Come on, let's go and help them hunt ...'

The door shut. Their footsteps died away.

The world spun in its little universe, about a foot across on the outside, infinitely large on the inside.

Behind it, stars floated away in the blackness. Here and there they congregated in great swirling masses, spinning about some unimaginable drain. Sometimes these drifted together, passing through one another like ghosts and parting in a trailing veil of stars.

Young stars grew in luminous cradles. Dead stars rolled in the glowing shrouds of their death.

Infinity unfolded. Walls of glittering swept past, revealing fresh fields of stars ...

... where, sailing through the endless night, made of hot gas and dust but recognizable nevertheless, was a turtle.

As above, so below.

INDEX